my **revisi⏻n** notes

CCEA GCSE

SCIENCE
SINGLE AWARD

James Napier

HODDER
EDUCATION
AN HACHETTE UK COMPANY

Although every effort has been made to ensure that website addresses are correct at time of going to press, Hodder Education cannot be held responsible for the content of any website mentioned in this book. It is sometimes possible to find a relocated web page by typing in the address of the home page for a website in the URL window of your browser.

Hachette UK's policy is to use papers that are natural, renewable and recyclable products and made from wood grown in well-managed forests and other controlled sources. The logging and manufacturing processes are expected to conform to the environmental regulations of the country of origin.

Orders: please contact Hachette UK Distribution, Hely Hutchinson Centre, Milton Road, Didcot, Oxfordshire, OX11 7HH. Telephone: +44 (0)1235 827827. Email education@hachette.co.uk Lines are open from 9 a.m. to 5 p.m., Monday to Friday. You can also order through our website: www.hoddereducation.co.uk

ISBN: 978 1 5104 0450 2

© James Napier 2017

First published in 2017 by
Hodder Education,
An Hachette UK Company
Carmelite House
50 Victoria Embankment
London
EC4Y 0DZ

www.hoddereducation.co.uk

Impression number   10 9

Year   2023

Cover photo © Panther Media GmbH/Alamy Stock Photo

Typeset in Bembo Std Regular 11/13 by Integra Software Services Pvt. Ltd., Pondicherry, India

Printed in India

A catalogue record for this title is available from the British Library.

Get the most from this book

Everyone has to decide his or her own revision strategy, but it is essential to review your work, learn it and test your understanding. These Revision Notes will help you to do that in a planned way, topic by topic. Use this book as the cornerstone of your revision and don't hesitate to write in it — personalise your notes and check your progress by ticking off each section as you revise.

Tick to track your progress

Use the revision planner on pages iv–vii to plan your revision, topic by topic. Tick each box when you have:

- revised and understood a topic
- tested yourself
- practised the exam questions and gone online to check your answers

You can also keep track of your revision by ticking off each topic heading in the book. You may find it helpful to add your own notes as you work through each topic.

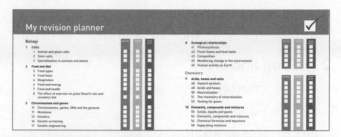

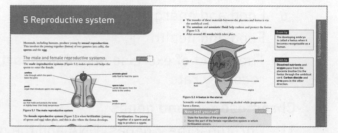

Features to help you succeed

My revision planner

Biology

1 Cells
  1 Animal and plant cells
  2 Stem cells
  3 Specialisation in animals and plants

2 Food and diet
  5 Food types
  5 Food tests
  6 Respiration
  6 Food and energy
  7 Food and health
  8 The effect of exercise on pulse (heart) rate and recovery rate

3 Chromosomes and genes
  11 Chromosomes, genes, DNA and the genome
  11 Mutations
  13 Genetics
  16 Genetic screening
  17 Genetic engineering

4 Coordination and control
  19 The nervous system
  20 Hormones

5 Reproductive system
  24 The male and female reproductive systems
  26 Contraception

6 Variation and adaptation
  30 Variation
  30 Natural selection

7 Disease and body defences
  34 Types of microorganism
  34 The body's defence mechanisms
  37 Antibiotics
  38 Development of medicines
  39 Alcohol and tobacco

REVISED   TESTED   EXAM READY

8 Ecological relationships

REVISED TESTED EXAM READY

41 Photosynthesis
43 Food chains and food webs
43 Competition
45 Monitoring change in the environment
45 Human activity on Earth

Chemistry

9 Acids, bases and salts

REVISED TESTED EXAM READY

48 Hazard symbols
48 Acids and bases
50 Neutralisation
51 The chemistry of neutralisation
53 Testing for gases

10 Elements, compounds and mixtures

55 Solids, liquids and gases
56 Elements, compounds and mixtures
56 Chemical formulae and equations
58 Separating mixtures

11 Periodic Table, atomic structure and bonding

63 The structure of the atom
66 The Periodic Table
66 Groups and periods
67 Compounds
68 The alkali metals (group 1)
68 The noble gases (group 0)
69 Ionic and covalent bonding

12 Metals and the reactivity series

72 The reactivity series
73 Energetics
74 Electrolysis
76 Flame tests

13 Materials

78 Natural and synthetic materials and their properties
79 Smart materials
79 Nanomaterials and emergent materials
81 Using materials to fight crime

14 Rates of reaction

84 Measuring the rate of reaction
86 Factors affecting the rate of reaction

15 Organic chemistry

88 Hydrocarbons
91 The alkenes
92 Atmospheric pollution
92 Polymers

Physics

16 Electrical circuits

95 Electrical circuits and symbols
97 Series and parallel circuits
99 Resistance

17 Household electricity

101 Protection from electrical shock
103 The cost of electricity

18 Energy

105 Energy transfers

19 Electricity generation

108 Making electricity
108 Power stations
109 Electricity transmission
110 Renewable and non-renewable sources of energy

20 Heat transfer

113 Methods of heat transfer
116 Heat transfer at the molecular level
117 Conserving heat in the home

21 Waves

119 Types of waves
120 Sound
123 The electromagnetic spectrum

REVISED   TESTED   EXAM READY

Exam practice answers at **www.hoddereducation.co.uk/myrevisionnotesdownloads**

22 Road transport and safety

126 Stopping a motor vehicle

127 Developing alternative fuels for transport and reducing reliance on fossil fuels

128 Road safety

129 Speed

130 Balanced and unbalanced forces

23 Radioactivity

133 Types of radiation

134 Half-life

135 Ionising radiation

24 Earth in space

137 The Solar System

138 Stars and galaxies

139 The expanding Universe

141 Glossary

150 Now test yourself answers

Exam practice answers at
www.hoddereducation.co.uk/myrevisionnotesdownloads

Countdown to my exams

3 weeks to go

- Start by looking at the specification — make sure you know exactly what material you need to revise and the style of the examination. Use the revision planner on pages iv–vii to familiarise yourself with the topics.
- Organise your notes, making sure you have covered everything on the specification. The revision planner will help you to group your notes into topics.
- Work out a realistic revision plan that will allow you time for relaxation. Set aside days and times for all the subjects that you need to study, and stick to your timetable.
- Set yourself sensible targets. Break your revision down into focused sessions of around 40 minutes, divided by breaks. These Revision Notes organise the basic facts into short, memorable sections to make revising easier.

REVISED ☐

2 weeks to go

- Read through the relevant sections of this book and refer to the exam tips and key terms. Tick off the topics as you feel confident about them. Highlight those topics you find difficult and look at them again in detail.
- Test your understanding of each topic by working through the 'Now test yourself' questions in the book. Look up the answers at the back of the book.
- Make a note of any problem areas as you revise, and ask your teacher to go over these in class.
- Look at past papers. They are one of the best ways to revise and practise your exam skills. Write or prepare planned answers to the exam practice questions provided in this book. Check your answers online at **www.hoddereducation/myrevisionnotesdownloads**
- Use the revision activities to try out different revision methods. For example, you can make notes using mind maps, spider diagrams or flash cards.
- Track your progress using the revision planner and give yourself a reward when you have achieved your target.

REVISED ☐

One week to go

- Try to fit in at least one more timed practice of an entire past paper and seek feedback from your teacher, comparing your work closely with the mark scheme.
- Check the revision planner to make sure you haven't missed out any topics. Brush up on any areas of difficulty by talking them over with a friend or getting help from your teacher.
- Attend any revision classes put on by your teacher. Remember, he or she is an expert at preparing people for examinations.

REVISED ☐

The day before the examination

- Flick through these Revision Notes for useful reminders, for example the exam tips and key terms.
- Check the time and place of your examination.
- Make sure you have everything you need — extra pens and pencils, tissues, a watch, bottled water.
- Allow some time to relax and have an early night to ensure you are fresh and alert for the examinations.

REVISED ☐

1 Cells

Animal and plant cells

Most animals and plants are formed of many millions of microscopic sub-units called **cells**.

Table 1.1 summarises the main features of animal and plant cells (as does Figure 1.1).

> **Cell**: The cell is the basic building block of all living organisms.

Table 1.1 Animal and plant cells

Structure	Function	Present in	
		animal cells	plant cells
Cell membrane	● Forms a boundary to the cell ● Selectively permeable, controlling what enters and leaves	Yes	Yes
Cytoplasm	● Site of chemical reactions	Yes	Yes
Nucleus	● Contains genetic information in the form of chromosomes ● The control centre of the cell	Yes	Yes
Cell wall	● Made of cellulose ● A rigid structure that provides support	No	Yes
(Large permanent) vacuole	● Contains cell sap ● Provides support	No	Yes
Chloroplasts	● Contain chlorophyll ● The place where photosynthesis takes place	No	Yes

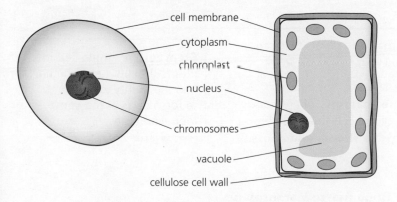

animal cell   plant cell

Figure 1.1 The structure of animal and plant cells

> **Exam tip**
>
> Both animal and plant cells contain a cell membrane, cytoplasm and a nucleus. Only plant cells contain a cellulose cell wall, vacuole and chloroplasts.

Prescribed practical B1

Carry out practical work to make a temporary slide and use a light microscope to examine and identify the structures of a typical plant and animal cell

Making a slide of a plant cell

● Peel a small section of onion tissue and place on the centre of a microscope slide.
● Add water using a drop pipette to the onion tissue to stop it drying out.
● Gently lower a coverslip onto the onion tissue. The coverslip will help protect the lens should the lens make contact with the slide and also prevent the cells drying out.
● Set the slide onto the stage of the microscope and examine using low power first and then high power.

Making a slide of an animal cell

● Using your nail or an inter-tooth brush, gently scrape the inside of your cheek.
● Smear the material gathered onto the centre of a microscope slide.
● Carefully lower a coverslip on top, as described for the plant cell.
● Observe using a light microscope, first at low power and then using high power.

Worked example

Mathematical skills can be tested in any part of the specification. For example, students could be asked to calculate the magnification of cells in a diagram provided suitable information is given.

The magnification of a microscope is the magnification of the eyepiece lens multiplied by the magnification of the objective lens being used.

A particular microscope has an eyepiece lens (×10) and two objective lenses (×10 and ×20).

What is the maximum magnification of this microscope?

Answer

200 (×10 multiplied by ×20)

Now test yourself

TESTED

1 Name **three** parts of a plant cell which are not found in animal cells.
2 Give the outermost part of a plant cell.
3 State **two** reasons for using a coverslip when observing cells under the microscope.

Answers on p. 150

Stem cells

REVISED

Stem cells are very simple cells found in animals and plants that can divide to form other cells of the same general type.

Stem cells are only found in certain parts of the body. For example, in humans stem cells can be found in the **bone marrow**. These bone marrow cells can make the different types of blood cell (but only blood cells). In plants, stem cells are found in the **tips of shoots and roots**.

> **Stem cell**: A simple cell in plants and animals which has the ability to divide to form cells of the same (general) type.

ⒽStem cells in medicine

Leukaemia is a type of cancer of the blood. **Bone marrow transplants** can be used as a form of treatment; the stem cells in the bone marrow from a donor contain the ability to produce the different types of blood cell in the right proportions (which doesn't happen in leukaemia).

Exam practice answers at **www.hoddereducation.co.uk/myrevisionnotesdownloads**

⊕Concerns over the use of stem cell research

- Stem cell research has **ethical implications**. Some people think that stem cell research using embryo stem cells could lead to 'designer babies'.
- **Viruses** or **diseases** could be transferred into the patient in the donor's bone marrow.
- People worry that **tumours** (cancer) or other **unwanted cell types** may develop.
- **Radiotherapy** or **chemotherapy** is often needed during pre-treatment to destroy the cancer cells before the bone marrow can be transplanted.

> **Exam tip**
>
> **Radiotherapy** is cancer treatment using high-energy X-rays or gamma radiation that can kill cells.
> **Chemotherapy** is treatment with special cancer drugs.

Specialisation in animals and plants

REVISED ☐

Stem cells in **multicellular** organisms (animals and plants) produce all the cell types needed. The levels of organisation are shown in Table 1.2.

Table 1.2 Cell specialisation

Structure	Description
Cell	Basic building block of living organisms, e.g. animal cell
Tissue	Group of cells with similar structures and functions, e.g. muscle
Organ	Group of different tissues working together to form a structure with a particular function, e.g. the brain
Organ system	Organs are organised into organ systems, e.g. the nervous system
Organism	The different organ systems make up the organism, e.g. a human

> **Exam tip**
>
> A multicelled organism contains many cells; humans contain many billions of cells.

> **Exam tip**
>
> In terms of complexity you should remember the order: cell – tissue – organ – organ system – organism.

Now test yourself

TESTED ☐

4 Name **one** disease that can be treated by using stem cells.
5 Define the term 'tissue'.

Answers on p. 150

Exam practice questions

1 (a) Figure 1.2 represents a plant cell.

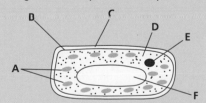

Figure 1.2

(i) Name the parts labelled **A**, **B** and **C**. [3]
(ii) Give the letters of **three** structures that are also present in animal cells. [3]

→

(b) Copy and complete the sentences below.

The _____ is the part of the cell that contains chromosomes.

Chemical reactions take place in the _____. [2]

2 (a) A thin piece of onion epidermis (skin) is placed onto a microscope slide. What else must you do before the slide is ready for observation using a microscope? [3]

(b) Suggest why onion cells do not contain chloroplasts. [1]

3 (a) Define the term 'stem cell'. [2]

(b) (i) Name **one** medical condition that can be treated using stem cells. [1]

(ii) Give **one** risk with using stem cells in medicine. [1]

(iii) Suggest why some people are ethically opposed to stem cell research. [1]

4 (a) Place the following terms in order of increasing complexity:

organ   tissue   cell   organ system [1]

(b) Define the term 'organ'. [1]

(c) Give **one** example of an organ in humans. [1]

Answers online

ONLINE

Exam practice answers at **www.hoddereducation.co.uk/myrevisionnotesdownloads**

2 Food and diet

We need a number of different **types** of food to keep us healthy. Food is also a source of **chemical energy**.

Food types

REVISED

Table 2.1 shows the main food types we need.

Table 2.1 The main food types (groups)

Food type	Needed in body for	Examples
Carbohydrate (starch)	Energy (slow-release)	Potato, bread
Carbohydrate (sugar)	Energy (fast-release)	Cake, biscuits
Protein	Growth and repair	Fish, beans
Fat	Energy store and provides insulation	Sausages, butter
Vitamin C	Healthy teeth and gums	Oranges, lemons
Vitamin D	Strong bones and teeth	Fish, milk
Calcium (mineral)	Strong bones and teeth	Milk, cheese
Iron (mineral)	Helping blood to carry oxygen	Red meat, spinach
Water	A solvent and for the transport of materials around the body	In all drinks and most foods
Fibre	Preventing constipation	Wholemeal bread, green vegetables

Food tests

REVISED

Food tests can be used to identify the food types present in different kinds of food.

Table 2.2 Food tests

Food type	Test	Method	Result (if food type present)
Starch	Starch test	Add iodine solution	Iodine turns from yellow-brown to blue-black
Sugar	Benedict's test	Add Benedict's reagent and **heat** in a water bath	The solution changes from blue to a brick-red precipitate
Protein/ amino acids	Biuret test	Add sodium hydroxide, then a few drops of copper sulfate and shake	The solution turns from a blue colour to a lilac/purple colour
Fats/oils	Ethanol test	Shake the fat/oil with alcohol, then add an equal amount of water	The clear colour of ethanol changes to a cloudy white emulsion

Exam tip

The Benedict's test is the only food test that requires heating.

Exam tip

Most foods contain more than one food type. Bacon, for example, contains protein, fat and water.

Respiration

Respiration is how we use carbohydrate (and sometimes fat) to provide **energy**. As respiration **releases energy**, it is an **exothermic** reaction.

The word equation for respiration is:

glucose + oxygen → carbon dioxide + water + energy

Ⓗ Higher-tier candidates need to know the balanced symbol equation for respiration:

$C_6H_{12}O_6 + 6O_2 \rightarrow 6CO_2 + 6H_2O$ + energy

> **Respiration**: The process that releases energy from food in all cells.
>
> **Exothermic**: A chemical reaction which releases energy (heat).

Food and energy

Prescribed practical B2

Investigate the energy content of food by burning food samples

The apparatus shown in Figure 2.1 can be used to compare the energy in different foods.

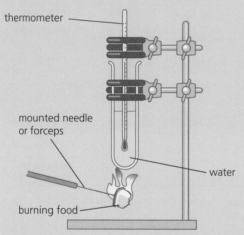

thermometer

mounted needle or forceps

water

burning food

Figure 2.1 Measuring the energy content of food

To make sure results are valid (a fair test) when comparing different foods:
- use the same amount of each food
- hold the burning food the same distance from the boiling tube
- use the same amount of water.

Some (heat) energy will be:
- lost to the air
- lost to heat the glass
- left in the burnt food remains.

Figure 2.2 Key things you must know in food-burning investigations

Exam tip

Many exam questions ask about **validity** and why all the energy in a food is not used to heat the water (i.e. why some is lost).

How much energy do we need?

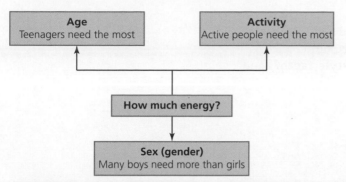

Figure 2.3 Factors that affect our energy needs

Now test yourself

TESTED

1 Name the food group that provides our main source of energy.
2 Which type of food is present if iodine turns from yellow-brown to blue-black?
3 Give the word equation for respiration.

Answers on p. 150

Food and health

REVISED

There is a clear link between diet and health.

There are a number of diseases and medical conditions that are a result of how much and what we eat. For example, **obesity** (being very overweight) is caused by eating too much carbohydrate and fat. **Circulatory diseases** are usually linked to the quality of our diet.

Circulatory diseases

Heart disease is an example of a circulatory disease (Figure 2.4).

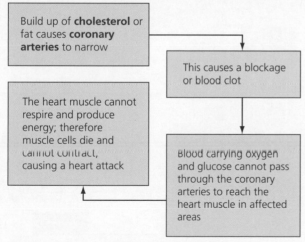

Figure 2.4 How heart attacks happen

Strokes are also examples of circulatory diseases, but they affect the **brain**.

Cholesterol: A fatty substance that causes narrowing of the blood vessels.

Coronary arteries: The blood vessels that supply the heart with blood.

Stroke: A circulatory disease that affects the brain.

How we can reduce the risk of having heart disease or strokes?

There are many factors that contribute to heart disease and strokes. Therefore, there are many things we can do to reduce our chances of getting heart disease; these can be grouped into **lifestyle factors** and **diet factors**, as shown in Figure 2.5.

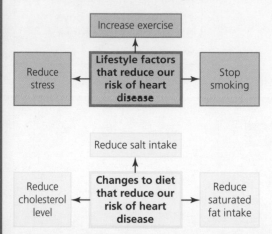

Figure 2.5 Lifestyle and diet factors that can help to reduce circulatory disease

However, many people fail to follow good advice about diet and lifestyle and remain at risk of heart attack or stroke.

Exam tip

You need to know the difference between lifestyle factors and diet factors.

The costs to society of circulatory diseases

Circulatory diseases have a big financial impact on both the NHS and affected families.

They are expensive to treat because:
- **many patients** are ill (in hospital) for a **long time**
- **expensive drugs and medicines** are often needed
- many **highly trained staff** are needed.

Families are affected because a parent may not be able to work or may require a lot of care if they have a circulatory disease.

Now test yourself

TESTED

4 Name the circulatory disease that affects the brain.
5 Name **three** lifestyle factors that can help protect against heart disease.

Answers on p. 150

The effect of exercise on pulse (heart) rate and recovery rate

REVISED

We can help our heart and help protect against heart disease by taking exercise.

Figure 2.6 shows the effect of exercise on the heart and recovery rates of two girls.

Exam practice answers at **www.hoddereducation.co.uk/myrevisionnotesdownloads**

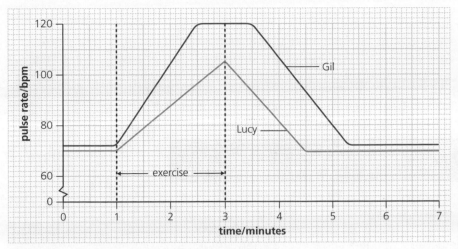

Figure 2.6 The effect of exercise on pulse rate

The **recovery time** is the length of time it takes after exercise for the pulse/heart rate to return to normal (the resting rate). A fit and healthy person will have a shorter recovery time than an unfit person.

Worked example

From Figure 2.6 state:
(a) when exercise started
(b) the maximum increase in pulse rate for Lucy
(c) the maximum percentage pulse rate increase for Lucy
(d) Lucy's recovery time.
(e) two things that suggest that Gil is less fit than Lucy.

Answer

(a) 1 minute
(b) 105 – 70 = 35
(c) $\frac{35}{70} \times 100 = 50\%$
(d) 3 minutes to 4.5 minutes = 1.5 minutes
(e) Any **two** from: Gil's resting rate is higher/her rate increases more during exercise/her recovery time is longer.

H Regular exercise **strengthens the heart muscle**. This increases the amount of blood pumped in each beat, even when at rest. Therefore the **cardiac output** (the amount of blood pumped per minute) is increased.

One consequence of having a stronger heart is that the heart will have to pump less often and, therefore, will suffer less wear and tear.

Exam practice questions

1 (a) Copy and complete the table below about some food tests. [3]

Reagent	Initial colour	End colour if food present
Iodine	Yellow-brown	
Biuret		Lilac/purple
	Clear	White emulsion

(b) Which food test requires heating to give a positive result? [1]

2 (a) Apart from energy, name **two** other products of respiration. [2]

(b) Respiration is an exothermic reaction. What does the term 'exothermic' mean? [1]

3 (a) Copy and complete the sentences below about food and health.
   There are a number of lifestyle changes we can make to reduce the risk of having a circulatory disease such as heart disease or _____. Apart from stopping _____ we can exercise more and reduce stress. [2]

   (b) State **two** dietary changes we can make to reduce our risk of circulatory disease. [2]

4 (a) Table 2.3 shows the pulse rate of two boys before, during and after exercise.

Table 2.3

Boy	Time/min									
	1	2	3	4	5	6	7	8	9	10
Jack	68	68	84	103	108	90	81	69	68	68
Sean	76	76	112	134	141	133	115	110	86	76

The boys started exercising after 2 minutes and stopped after 5 minutes.

   (i) Calculate the maximum increase in Jack's heart rate compared with the resting rate. [1]

   (ii) How long did it take Sean's pulse rate to return to the normal resting rate after the exercise stopped? [1]

   (iii) Give **two** pieces of evidence that suggest that Jack is the fitter of the two boys. [2]

   (b) (i) Define the term 'cardiac output'. [1]

   (ii) Exercise results in an increased cardiac output at rest. Explain the advantage of this to the heart. [2]

Answers online

ONLINE

3 Chromosomes and genes

Chromosomes are genetic structures found in the nucleus of animal and plant cells.

Chromosomes, genes, DNA and the genome

Figure 3.1 shows that a chromosome is divided into many smaller units called **genes**.

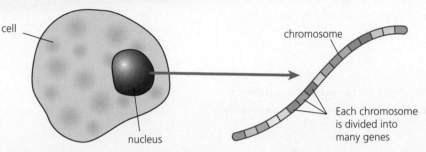

Figure 3.1 Chromosomes and genes

Each gene controls a particular characteristic (feature) such as eye colour.

The chromosomes are made up of the chemical **DNA**, which is formed into a **double helix** shape (Figure 3.2).

Chromosome: A genetic structure that occurs in functional pairs in the nucleus of cells (except gametes, where there is only one chromosome from each pair).

Gene: A short section of a chromosome that codes for a particular characteristic.

DNA: The molecule (core component) that forms genes and chromosomes.

Double helix: The structure of DNA.

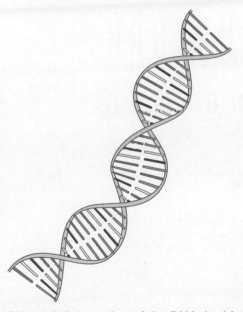

Figure 3.2 A section of the DNA double helix

All the DNA in an organism (all the genetic material) is referred to as its **genome**.

Genome: The entire genetic (DNA) make-up of an individual.

Exam tip

DNA is the name of the molecule (the core component) that makes genes and chromosomes; the double helix refers to its shape or structure.

Mutations

Mutations are random **changes** in the structure of genes or number of chromosomes. Mutations in the cells of living organisms can often be harmful.

Mutation: A random change in the structure of a gene or number of chromosomes.

Cancer

When an organism grows, this usually involves it producing more cells. In this process, cells divide (split) into two and then each of the two new cells grows to normal size and eventually divides again, and so on.

Cancer is **uncontrolled cell division** caused by damage (change) to the genes or chromosomes in a part of the body. Cancer is caused by mutations.

Some mutations (cancers) can be triggered by **environmental factors**. **Ultraviolet (UV) light** coming from the Sun can cause mutations in skin cells leading to **skin cancer**.

Cancer: Uncontrolled cell division.

Genetic conditions

A **genetic condition** is a medical condition caused by mutations involving chromosomes or genes.

Down's Syndrome

Down's Syndrome is caused by a change in **chromosome number**. Individuals with Down's Syndrome have 47 rather than 46 chromosomes in each cell.

It is possible to tell if someone has Down's Syndrome by studying a **karyotype**. This is a diagram or photograph of all the chromosomes in a cell carefully laid out (usually in pairs) so that they can be counted (Figure 3.3).

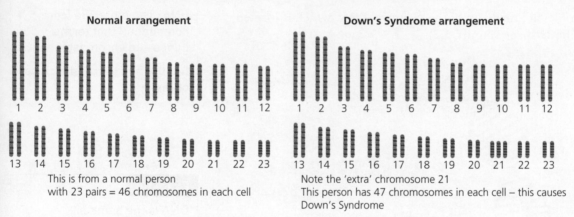

Normal arrangement

1 2 3 4 5 6 7 8 9 10 11 12
13 14 15 16 17 18 19 20 21 22 23

This is from a normal person with 23 pairs = 46 chromosomes in each cell

Down's Syndrome arrangement

1 2 3 4 5 6 7 8 9 10 11 12
13 14 15 16 17 18 19 20 21 22 23

Note the 'extra' chromosome 21 This person has 47 chromosomes in each cell – this causes Down's Syndrome

Figure 3.3 Karyotypes showing different chromosome arrangements

Cystic fibrosis

Unlike Down's Syndrome, which is caused by an alteration in chromosome number, individuals have cystic fibrosis due to **gene mutations** being inherited from parents at birth. It is an example of an **inherited** disease.

Now test yourself

TESTED

1 Name the part of the cell that contains DNA.
2 What term is used to describe the shape of DNA?
3 Name the genetic condition caused by having one chromosome too many.

Answers on p. 150

⊕Genetics

Genetics explains how characteristics (features) such as eye colour pass from parents to offspring. It is the information in the genes on the chromosomes that does this.

Chromosomes are arranged in pairs; in any one pair the two chromosomes carry the same genes (for example, both chromosomes carry genes for eye colour). However, the form of the gene (such as blue or brown eyes) in the two partner chromosomes can be different.

If a gene exists in two forms, each form is called an **allele**. If the two alleles of a gene are the same (in one individual/pair of chromosomes), they are **homozygous**; if they are different, they are **heterozygous** (Figure 3.4).

> **Allele**: One of two possible versions of a particular gene.
>
> **Homozygous**: Both alleles of a particular gene are the same.
>
> **Heterozygous**: The two alleles of a particular gene are different.

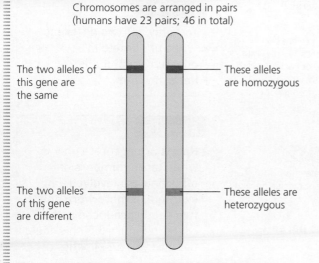

Chromosomes are arranged in pairs (humans have 23 pairs; 46 in total)

The two alleles of this gene are the same — These alleles are homozygous

The two alleles of this gene are different — These alleles are heterozygous

Figure 3.4 Chromosomes, genes and alleles

Genetic diagrams

Genetics questions in *Science: Single Award* cover only the variation produced by one gene and its pair of alleles.

Some key features about genetic crosses:
- The crosses are always about **one characteristic**, such as eye colour, flower colour, height in peas.
- **Gametes** (the sex cells (e.g. sperm and eggs) produced by individuals) will contain one chromosome and therefore only one allele from each pair of chromosomes. The two chromosomes of each pair then recombine following fertilisation in sexual reproduction.
- The **phenotype** is the outward appearance of an individual or feature, for example blue eyes.
- The **genotype** is a set of paired symbols representing the two alleles in an individual.
- In a heterozygous individual, the **dominant allele** overrides the non-dominant **recessive allele**. (In other words, it is the dominant allele that will be shown in the phenotype.)
- The recessive allele is dominated by the dominant allele. The recessive feature only shows itself in the phenotype if there are two recessive alleles (and no dominant allele).
- When setting out a genetic cross it is usually in the following sequence: parents' phenotypes; parents' genotypes; gametes; offspring; offspring genotypes; offspring phenotypes. This is illustrated in Figure 3.5.

> **Phenotype**: The outward appearance of a feature.
>
> **Genotype**: The genetic make-up of an individual represented by two symbols (letters).
>
> **Dominant allele**: In the heterozygous condition, the dominant allele overrides the recessive allele.
>
> **Recessive allele**: An allele that will only show in the phenotype if both alleles are recessive (and there is no dominant allele present).

Examples of genetic crosses

In pea plants, seeds are either round or wrinkled. Seed shape is controlled by one gene (on each of two partner chromosomes) but there can be two different alleles. The allele for round seed is dominant to the (recessive) allele for wrinkled seed. The alleles are given the symbols R (for round) and r (for wrinkled).

Figure 3.5 shows the offspring genotypes and phenotypes produced from a cross between two pea plants, each heterozygous for seed shape.

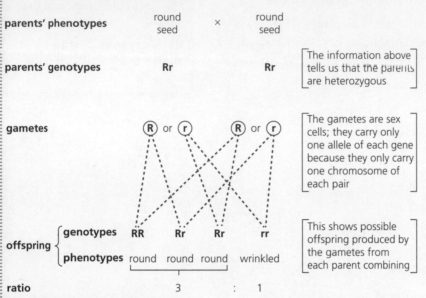

Figure 3.5 A genetic cross showing how two heterozygous parents produce offspring in a 3 : 1 ratio.

Exam tip

The gametes produced by one parent can combine only with the gametes of another parent; the different gametes of the same individual cannot combine.

Exam tip

You only get two different types of gamete in one individual if it is heterozygous.

Exam tip

Ratios are accurate only when large numbers of offspring are involved. For example, if there were only two seeds produced in the cross in Figure 3.5, this could not give you a 3 : 1 ratio.

It is normally easier to use **Punnett squares** when setting out genetic crosses, as shown in Figure 3.6. In this example, using seed shape in peas as before, a heterozygote (**Rr**) pea is crossed with a homozygous recessive (**rr**) pea.

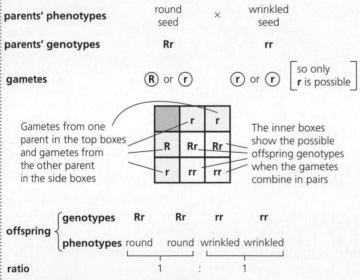

Figure 3.6 Using a Punnett square

Exam tip

If the offspring ratio is 3 : 1 in a genetic cross, then **both parents** must have been **heterozygous**.

If the offspring ratio is 1 : 1, then one parent is **heterozygous** and the other is **homozygous recessive**.

Exam practice answers at **www.hoddereducation.co.uk/myrevisionnotesdownloads**

Worked example

Brown eyes are dominant to blue eyes. Using the symbols **B** for brown and **b** for blue, draw a Punnett square to show how brown-eyed parents can have children with blue eyes.

Answer

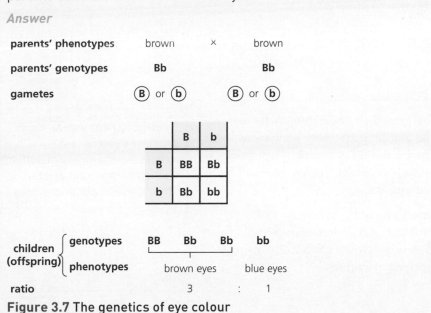

parents' phenotypes	brown	×	brown
parents' genotypes	**Bb**		**Bb**
gametes	Ⓑ or Ⓑ		Ⓑ or Ⓑ

	B	b
B	BB	Bb
b	Bb	bb

children (offspring)
- genotypes: **BB Bb Bb bb**
- phenotypes: brown eyes | blue eyes
- ratio: 3 : 1

Figure 3.7 The genetics of eye colour

Students need to understand the terms **ratio**, **percentage** and **probability**.

Worked example

The allele for cystic fibrosis is recessive to the normal allele. A cross between two parents who are each heterozygous for cystic fibrosis gives the offspring shown in the Punnett square in Figure 3.8 (**F** = normal allele; **f** = cystic fibrosis allele).

	F	f
F	FF	Ff
f	Ff	ff

Figure 3.8 Cross between two heterozygous parents, each carrying the gene for cystic fibrosis

(a) Give the ratio of offspring phenotypes produced.
(b) What percentage of the offspring are homozygous?
(c) What is the probability of the next child to these parents having cystic fibrosis?

Answer

(a) The **ratio** of offspring produced is three normal to one cystic fibrosis.
(b) The **percentage** of the offspring that are homozygous (**FF** and **ff**) is 50%.
(c) The **probability** of their next child having cystic fibrosis is 1 in 4 or 1 : 3.

Genetic questions can also be asked in the form of **pedigree diagrams**. A pedigree diagram shows how a genetic condition passes down through different members of the same family. There is an example of a pedigree diagram in the Exam practice questions at the end of this chapter.

> **Pedigree diagram**: A diagram that shows how a particular condition is inherited through the different generations of a family.

Now test yourself

TESTED

4 In terms of alleles, explain what is meant by the term 'homozygous'.
5 Define the term 'recessive'.
6 How many alleles of a gene are found in a gamete?

Answers on p. 150

Genetic screening

REVISED

Genetic screening means **testing for genetic conditions**. A foetus can be screened for certain conditions but there are other conditions for which the parents can be tested before becoming pregnant. This type of screening checks whether the parents have alleles that could result in them having a child with an inherited condition.

Down's Syndrome is a human condition caused by having an extra chromosome (47 rather than 46). For many years it has been possible to test a foetus to check whether it will develop Down's Syndrome using a process called **amniocentesis**, as described in Figure 3.9.

> **Genetic screening**: A process used to test people for the presence of particular harmful alleles or other genetic abnormalities.

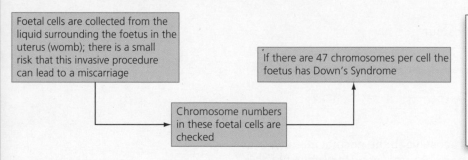

Foetal cells are collected from the liquid surrounding the foetus in the uterus (womb); there is a small risk that this invasive procedure can lead to a miscarriage

If there are 47 chromosomes per cell the foetus has Down's Syndrome

Chromosome numbers in these foetal cells are checked

> **Amniocentesis**: A process in which foetal cells are obtained from the amniotic fluid and then examined for the presence of genetic abnormalities.

Figure 3.9 Using amniocentesis to test for Down's Syndrome

In recent years a test has been developed that can indicate whether a foetus may have Down's Syndrome using a blood sample from the pregnant mother. Blood testing is normally not as accurate as amniocentesis but doesn't carry the risk of miscarriage.

> **Exam tip**
>
> Blood testing is quicker, less invasive and less risky, but is usually not as accurate as amniocentesis.

Genetic screening: ethical issues

Many genetic conditions in foetuses, and also in adults, can now be genetically screened. However, genetic screening has many ethical issues.
- Should you be allowed a free choice of whether to screen or not? Who should really make the decision?
- Should 'cost' be a factor?
- Who should be allowed access to the information? Should it be available to insurance companies and employers? If insurance companies have genetic information that someone is more likely to die young or become disabled, will insurance be more difficult to get and will it be more expensive?

The parents of a foetus that has been screened to show that it has a genetic condition that will have a very significant impact on its quality of life face a very difficult **ethical dilemma**. The parents will have to choose whether to have an abortion or to have a child that will need constant care and may have a very poor quality of life. However, many parents are against abortion for ethical or moral reasons, and in some countries it is not allowed.

> **Exam tip**
>
> Because of the risk of miscarriage with amniocentesis, this process is usually recommended only for older mothers where the risk of Down's Syndrome is higher (or where other evidence suggests that there is a raised chance that the foetus has Down's Syndrome).

Exam practice answers at **www.hoddereducation.co.uk/myrevisionnotesdownloads**

⊕Genetic engineering

Genetic engineering is a process which modifies (changes) the genome (DNA) of an organism.

Genetic engineering normally involves taking a piece of DNA–usually a gene–from one organism (the donor) and adding it to the genetic material of another organism (the recipient).

Commonly, DNA for a desired product (such as a human hormone) is incorporated into the DNA of bacteria. This is because bacterial DNA is easily manipulated and also because bacteria reproduce so rapidly that large numbers containing the new gene can be produced quickly.

As with many recently developed technologies, there are advantages and disadvantages with genetic engineering.

Advantages:

● **Large quantities** of **human insulin** can be produced. Before genetic engineering, people with diabetes had to use insulin extracted from pigs and cattle in abattoirs; usually only small amounts could be extracted and the animal insulin was **slightly different** in structure from human insulin.

Disadvantages:

● Many people have **moral issues** about altering an organism's DNA and adding the DNA from one organism to another.
● There could be unexpected or unforeseen outcomes as a result of the technique.
● Some of the DNA (genes) being transferred could make it into the environment (wild) accidentally.

> **Genetic engineering**: A process which modifies the genome of an organism.

> **Exam tip**
>
> Genetically modified (GM) crops are produced by genetic engineering. Advantages include higher crop yields and disease-resistant varieties. A potential disadvantage is the development of 'super-weeds' caused by the spread of genes from the GM crops to common weed species.

Now test yourself

7 Define the term 'genetic screening'.
⊕ 8 Give **one** advantage of genetic engineering.

Answers on p. 150

Exam practice questions

1 Place the following structures in order of size, starting with the smallest:
chromosome   cell   gene   nucleus   [1]
2 (a) (i) Explain the term 'mutation'.   [2]
      (ii) Describe fully how skin cancer is caused.   [2]
  (b) Cystic fibrosis and Down's Syndrome are genetic conditions caused by mutations.
      (i) Name the type of mutation that leads to cystic fibrosis.   [1]
      (ii) Copy and complete the sentence below:
           Normally we have 46 chromosomes in each cell in the body but individuals with
           Down's Syndrome have _____ chromosomes.   [1]
⊕ 3 Pea plants can be either tall or short. When two tall plants were crossed, three quarters of the offspring were tall and the rest were short. The tall allele (**T**) is dominant to the short allele (**t**).
  (a) Copy and complete the Punnett square in Figure 3.10 to explain the outcome of this cross. [2]

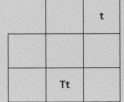

Figure 3.10

(b) (i)   What percentage of the offspring are heterozygous? [1]

(ii) Give the genotype of offspring that are homozygous recessive. [1]

4 Huntington's disease is a rare medical condition caused by the presence of a single allele. The pedigree diagram shows the inheritance of Huntington's disease in a family through three generations.

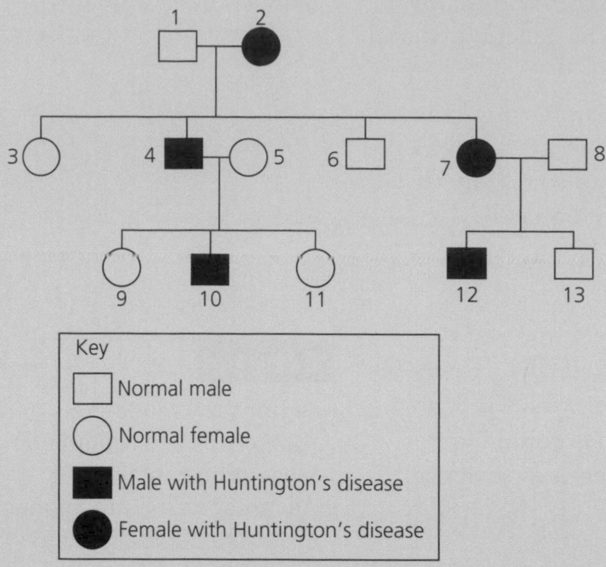

Key
☐ Normal male
○ Normal female
■ Male with Huntington's disease
● Female with Huntington's disease

Figure 3.11 The inheritance of Huntington's disease

(a) How many female grandchildren do individuals 1 and 2 have? [1]

(b) What is the evidence that the allele for Huntington's disease is dominant and not recessive? [1]

5 (a) (i)   Give **one** example of a medical condition that can be identified by the genetic screening of a foetus. [1]

(ii) Describe the process of amniocentesis. [3]

(b) It is now possible to screen adults for the presence of harmful alleles that could cause illness when they are older.

Suggest **one** argument against making this type of information public. [1]

6 (a) Define the term 'genetic engineering'. [1]

(b) Bacteria (and other microorganisms) can be modified to produce products of value to man.

(i)   Name **one** product produced by this method. [1]

(ii) Give **two** possible disadvantages of genetic engineering. [2]

Answers online

ONLINE ☐

4 Coordination and control

The nervous system

We are able to respond to the environment around us. Anything that we respond to is called a **stimulus** (plural, **stimuli**).

In animals, stimuli (which can include sounds, smells or visual stimuli) affect **receptors** in the body. There are many types of receptor, each responding to a particular type of stimulus. If a receptor is stimulated, it may cause an **effector**, such as a **muscle**, to produce a **response**.

The role of the nervous system in animals is summarised in Figure 4.1.

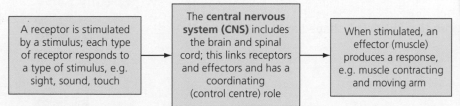

Figure 4.1 The nervous system

Central nervous system (CNS): The part of the nervous system that links receptors and effectors, comprising the brain and the spinal cord.

Exam tip

Remember that the CNS includes both the brain and the spinal cord.

Voluntary and reflex actions

Voluntary actions and **reflex actions** are the two main types of nervous action. They are summarised in Table 4.1.

Table 4.1 Voluntary and reflex actions

	Voluntary	Reflex
Thinking time (conscious control) involved	Yes	No
Speed of response	Variable – usually much slower	Fast

Voluntary action: An action or response that involves conscious thought.

Reflex action: A rapid, involuntary action that does not involve conscious thought.

Exam tip

Reflexes are **automatic** and often **protective**, such as the withdrawal of a hand from a hot object.

H The pathway of neurones in a reflex action is described as a **reflex arc**. The reflex arc for the reflex that occurs when someone puts his or her hand on a hot object is shown in Figure 4.2. The gaps between neurones are called **synapses**.

Synapse: A small gap between neurones.

association (connector) neurone
joins the sensory and motor neurones

sensory neurone
carries nerve impulses from the receptors, for example a burn on the hand, to the spinal cord

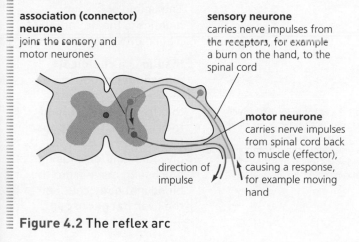

motor neurone
carries nerve impulses from spinal cord back to muscle (effector), causing a response, for example moving hand

direction of impulse

Figure 4.2 The reflex arc

H **Worked example**

Students could be asked to draw or complete a diagram of a reflex arc. Figure 4.3 shows a typical response from a student when asked to draw a reflex arc. Can you identify the mistakes in this diagram?

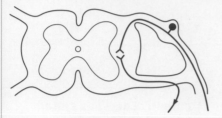

Figure 4.3 An incorrect diagram of a reflex arc

Answer

- The association neurone is missing.
- The motor neurone should continue along the gap before dropping (alongside the sensory neurone).
- The neurones should extend into the 'butterfly'-shaped area in the centre of the spinal cord.

Now test yourself

1 Name the **two** parts of the CNS.
2 Define the term 'reflex action'.
3 Name the neurone that links the sensory and motor neurones in a reflex arc.

Answers on p. 150

TESTED ☐

Hormones

REVISED ☐

Hormones are **chemical messengers produced by glands** that travel in the **blood** to bring about a response in a **target organ.**

The main differences between hormones and the nervous system are summarised in Table 4.2.

Table 4.2 The main differences between hormones and nervous communication

	Nervous system	Hormones
Method of communication	Impulses along neurones	Chemicals in blood
Speed of action	Fast-acting	Usually slow-acting

> **Hormone**: A chemical messenger produced by a gland that travels in the blood to a target organ where it acts.

> **Exam tip**
>
> In an exam you need to state that insulin lowers/reduces blood glucose levels. It is not enough to say that it controls blood glucose levels.

Insulin

Insulin is a hormone that **lowers blood glucose** levels (Figure 4.4). It is important that the amount of glucose (sugar) in the blood is at just the right level.

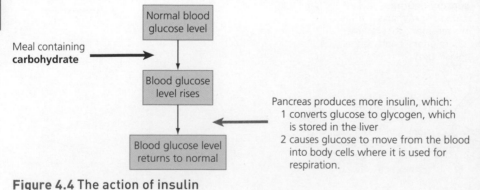

Figure 4.4 The action of insulin

> **Exam tip**
>
> The action of insulin highlights the definition of a hormone.
>
> **Insulin** is a **chemical messenger** produced by the pancreas (**a gland**) that travels in the **blood** to bring about a **response** (converting glucose to glycogen and/or causing glucose to move from the blood into body cells) in the liver (**target organ**).

> **Worked example**
>
> (a) Explain why the body needs glucose in the blood.
> (b) Suggest why the level of insulin in the blood is often at its lowest in the middle of the night.
>
> *Answer*
>
> (a) The glucose is transported to body cells to release energy by respiration.
> (b) Glucose levels will also be low during the night. By this time the glucose from the last meal of the day has been used up in respiration or converted to glycogen for storage.

Diabetes

Diabetes is a lifelong condition in which the body does not produce enough insulin (or the insulin produced does not work). People with diabetes can have very high (and dangerous) blood glucose levels, unless treated.

The **symptoms** (signs that show something is wrong) of diabetes include:

● high blood glucose levels
● glucose in the urine
● lethargy
● thirst.

The main differences between Type 1 and Type 2 diabetes are summarised in Table 4.3.

> **Diabetes**: A condition in which the blood glucose control mechanism fails.

Table 4.3 The main differences between Type 1 and Type 2 diabetes

	Type 1	Type 2
Main effect	Insulin is not produced by the pancreas	Insulin is produced but stops working properly or the pancreas does not produce enough insulin
Treatment	Insulin injections (plus controlled diet and exercise)	Usually controlled by diet initially, but later requires medication and/or insulin injections
Preventative measures	None – not caused by lifestyle	Take exercise, reduce sugar intake, avoid obesity
Age of first occurrence	Often in childhood	Usually as an adult

Long-term effects and future trends

People who have had diabetes for a long time and or whose blood glucose level is not tightly controlled are at risk of developing **long-term complications**. These include:

● **eye damage** (and blindness)
● **heart disease** and **strokes** (circulatory diseases)
● **kidney damage**.

The number of people who suffer from diabetes is increasing rapidly and the cost of treatment is becoming very high. The large increase in the number of people with Type 2 diabetes is linked to poor diet and a lack of exercise.

> **Exam tip**
>
> Type 2 diabetes is linked to lifestyle but Type 1 diabetes is *not* caused by lifestyle.

Now test yourself

TESTED ☐

4 Name the organ that produces insulin.
5 Give **two** ways in which insulin lowers blood glucose levels.
6 Give **three** symptoms of diabetes.

Answers on p. 150

Plant hormones

Hormones are also important in coordination in plants. **Phototropism** is the growth response in which plants bend in the direction of light.

> **Exam tip**
>
> By bending in the direction of the light, the plant gets more light and so more photosynthesis and more growth take place.

> **Phototropism**: A plant growth response which results in plant stems growing in the direction of a light source.

H The plant hormone that causes phototropism is **auxin**. This hormone stimulates growth in cells. Figure 4.5 shows that the auxin is produced in the tip of the plant but more passes down the shaded (non-illuminated side) than the illuminated side. This means that the cells in this region get more auxin and therefore grow more (compared with the cells in the side getting most light).

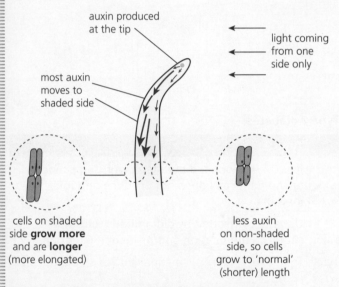

- auxin produced at the tip
- most auxin moves to shaded side
- light coming from one side only
- cells on shaded side **grow more** and are **longer** (more elongated)
- less auxin on non-shaded side, so cells grow to 'normal' (shorter) length

Figure 4.5 The role of auxin in phototropism

Exam practice questions

1 (a) Name the two parts of the CNS. [2]

  (b) Copy and complete Table 4.4 to show the differences between voluntary and reflex actions. [2]

   Table 4.4

	Voluntary	Reflex
Conscious control (thinking time involved)		
Speed of action		

2 (a) Copy and complete the sentences below about hormones.

   A hormone is a chemical _____ produced by glands that travels in the _____. The hormones are transported to a _____ organ, where they act. [3]

  (b) (i) Name the organ in which insulin is produced. [1]

   (ii) What is the function of insulin? [1]

   (iii) Describe **two** ways in which it carries out this function. [2]

Exam practice answers at **www.hoddereducation.co.uk/myrevisionnotesdownloads**

3 (a) Copy and complete Table 4.5 by putting the symptoms and long-term effects of diabetes in the correct columns. Choose from: eye damage, glucose in the urine, thirst, stroke. [2]

Table 4.5

Symptom	Long-term effect (complication)

(b) Suggest **three** reasons why the cost to the NHS of treating diabetes is so high. [3]

4 (a) Figure 4.6 shows some seedlings that have light coming from one side only.

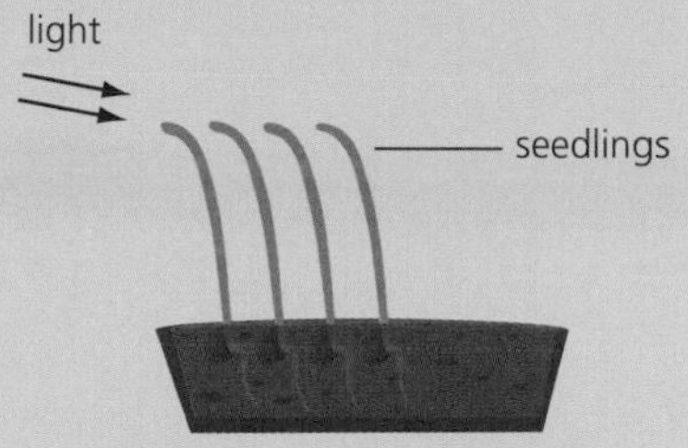

Figure 4.6

(i) Name this growth response in plants. [1]
(ii) Explain how this response benefits the seedlings. [2]

(b)

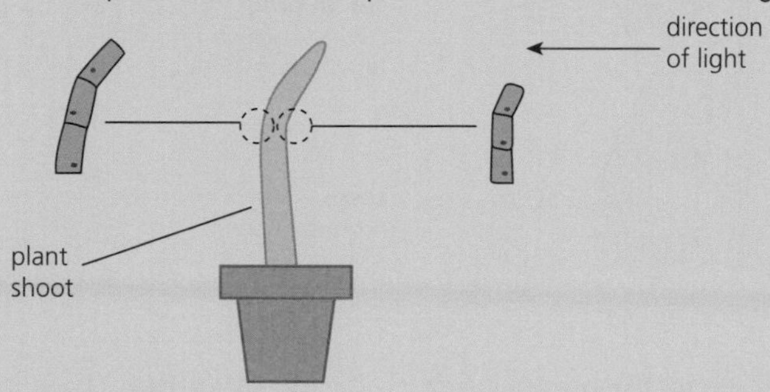

Figure 4.7

(i) Name the plant hormone involved in this response. [1]
(ii) Use Figure 4.7 and your knowledge to explain why this plant shoot is bending to the right. [3]

Answers online

ONLINE

5 Reproductive system

Mammals, including humans, produce young by **sexual reproduction**. This involves the joining together (fusion) of two gametes (sex cells), the **sperm** and the **egg**.

The male and female reproductive systems

REVISED

The **male reproductive system** (Figure 5.1) makes sperm and helps the sperm to enter the female.

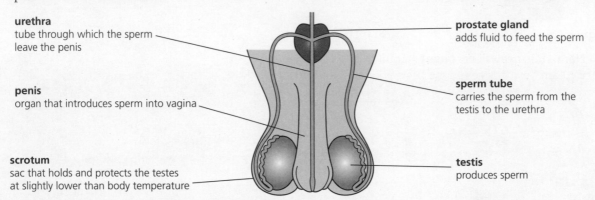

urethra
tube through which the sperm leave the penis

penis
organ that introduces sperm into vagina

scrotum
sac that holds and protects the testes at slightly lower than body temperature

prostate gland
adds fluid to feed the sperm

sperm tube
carries the sperm from the testis to the urethra

testis
produces sperm

Figure 5.1 The male reproductive system

The **female reproductive system** (Figure 5.2) is where **fertilisation** (joining of sperm and egg) takes place, and this is also where the foetus develops.

> **Fertilisation**: The joining together of a sperm and an egg to produce a zygote.

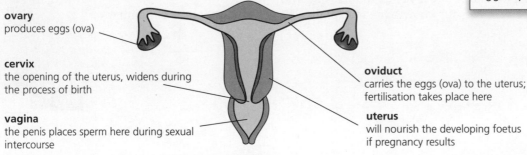

ovary
produces eggs (ova)

cervix
the opening of the uterus, widens during the process of birth

vagina
the penis places sperm here during sexual intercourse

oviduct
carries the eggs (ova) to the uterus; fertilisation takes place here

uterus
will nourish the developing foetus if pregnancy results

Figure 5.2 The female reproductive system

Fertilisation and pregnancy

If a sperm and an egg meet and **fuse** (join) in an **oviduct**, fertilisation will result.

The fertilised egg becomes the first cell (**zygote**) of the new individual.

The stages that follow fertilisation (**pregnancy**) are described below:
- The zygote divides many times to form a ball of cells as it travels down the oviduct to the uterus.
- This ball of cells (embryo) **implants** in the uterus lining (**implantation**).
- The **placenta** develops and is linked to the foetus by the **umbilical cord**.
- The placenta is where the **exchange of dissolved nutrients** (e.g. glucose), **oxygen**, **carbon dioxide** and **urea** (a waste product) occurs between the mother and the foetus.

> **Zygote**: The first (diploid) cell of the new individual following fertilisation.
>
> **Implantation**: The term describing the attachment of the ball of cells (embryo) following fertilisation to the uterus lining.

- The transfer of these materials between the placenta and foetus is via the umbilical cord.
- The **amnion** and **amniotic fluid** help cushion and protect the foetus (Figure 5.3).
- After around **40 weeks** birth takes place.

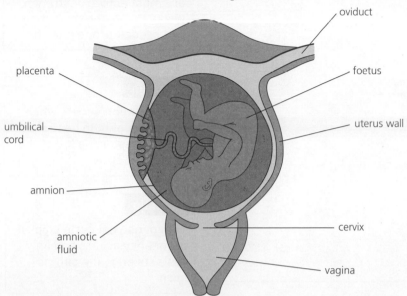

oviduct

placenta

foetus

umbilical cord

uterus wall

amnion

amniotic fluid

cervix

vagina

Figure 5.3 A foetus in the uterus

> **Exam tip**
>
> The developing embryo is called a foetus when it becomes recognisable as a human.

> **Exam tip**
>
> **Dissolved nutrients** and **oxygen** pass from the placenta (mother) to the foetus through the umbilical cord. **Carbon dioxide** and **urea** pass in the other direction.

Scientific evidence shows that consuming alcohol while pregnant can harm a foetus.

Now test yourself

TESTED

1 State the function of the prostate gland in males.
2 Name the part of the female reproductive system in which fertilisation occurs.
3 What is the function of the amniotic fluid during pregnancy?

Answers on p. 150

Ⓗ The normal number of chromosomes in a cell is described as the **diploid** number (e.g. 46 in humans). Sperm and egg cells (nuclei) only contain half the normal number of chromosomes (e.g. 23 in humans). They are described as being **haploid**. Fertilisation involves a haploid sperm and a haploid egg fusing so that the zygote has the diploid (normal chromosome) number (46 in humans).

The scientific name for an egg is **ovum**.

The menstrual cycle

The process of **menstruation** (having periods) starts in girls at puberty and continues until the end of a woman's reproductive life. The function of the **menstrual cycle** is the monthly renewal of the delicate blood-rich lining of the uterus so that it will provide a suitable environment for the embryo should fertilisation occur.

The menstrual cycle lasts (approximately) 28 days. The ovum is released (**ovulation**) on around day 14. By this time the uterine lining has built up in preparation for pregnancy.

Sexual reproduction can result in pregnancy if it occurs in a short window on either side of ovulation. Menstruation is the breakdown and removal of the blood-rich uterine lining at the end of each cycle (Figure 5.4).

> **Diploid**: Describes a cell or nucleus with the normal chromosome number.
>
> **Haploid**: Describes a cell or nucleus with half the normal number of chromosomes.

> **Exam tip**
>
> It is important that **all gametes are haploid**. If they weren't, then every time fertilisation took place, the chromosome numbers in a cell would double.

> **Menstruation**: The breakdown and removal of the blood-rich uterine lining at the end of each cycle.
>
> **Ovulation**: The release of an ovum (egg) by an ovary.

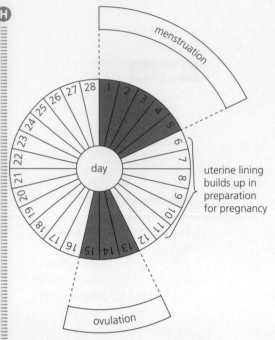

Figure 5.4 The menstrual cycle

uterine lining builds up in preparation for pregnancy

The menstrual cycle is controlled by **hormones** including:

● **oestrogen** – this stimulates ovulation and starts the build-up and repair of the uterine wall after menstruation
● **progesterone** – this continues the build-up of the uterine lining after ovulation.

Now test yourself

TESTED

4 Name **two** cells that are haploid.
5 Define the term 'menstruation'.

Answers on p. 150

Contraception

REVISED

Contraception reduces (or stops) the possibility of pregnancy occurring when having sex.

Some people who are opposed to contraception for **ethical** or **moral** reasons can reduce the chances of pregnancy by avoiding having sex around the time of ovulation each month.

The three main types of contraception are:
● mechanical (physical)
● chemical
● surgical.

Examples of each type of contraception and an explanation of how they work, together with their main advantages and disadvantages, are given in Table 5.1.

Table 5.1 Methods of contraception

Type	Example	Method	Advantages	Disadvantages
Mechanical (physical)	Male and female condom	Acts as a barrier to trap sperm and prevent them swimming up the female's reproductive system	● Easily obtained and also protects against sexually transmitted infections such as HIV (leading to AIDS) and chlamydia ● Some STIs can lead to infertility if untreated, e.g. chlamydia	● Unreliable if not used properly
Chemical	Contraceptive pill	Taken regularly by the woman and prevents the ovaries releasing eggs by changing hormone levels	● Very reliable	● Can cause some side effects such as weight gain and mood swings, and may increase the risk of blood clots ● The woman needs to remember to take the pill daily for around 21 consecutive days in each cycle
	Implant	An implant is a small tube about 4 cm long that is inserted just under the skin in the arm and releases hormones slowly over a long period of time	● Very reliable ● Can work for up to 3 years	● Do not protect against STIs ● Can prevent menstruation taking place
Surgical	Vasectomy (male sterilisation)	Cutting of sperm tubes, preventing sperm from entering the penis	● Virtually 100% reliable	● Very difficult or impossible to reverse ● Does not protect against STIs
	Female sterilisation	Cutting of oviducts, preventing eggs from moving through the oviduct and being fertilised	● Virtually 100% reliable	● Very difficult or impossible to reverse ● Does not protect against STIs

You need to be able to compare the advantages and disadvantages of each type of contraception and be able to decide which methods might be better in any particular situation.

Worked example

Suggest why condoms are a widely used contraceptive for young married couples who have no children as yet.

Answer

Condoms are not permanent (unlike sterilisation) and the couples may hope to have children later.

Condoms have no side effects, unlike the contraceptive pill. They are also easily obtained.

Exam practice questions

1 (a) Figure 5.5 represents the female reproductive system.

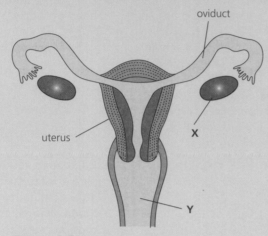

Figure 5.5

   (i) Name parts **X** and **Y**. [2]
   (ii) Name the place where fertilisation takes place. [1]
   (iii) State **one** change that takes place in the uterus during pregnancy. [1]
 (b) Describe the passage of sperm from where it is produced in the testes to where it leaves the male
   body. Your answer should include all the structures, in order, that it passes through. [3]

2 Figure 5.6 shows the placenta and surrounding structures.

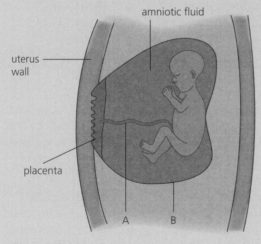

Figure 5.6

 (a) Identify structure **A**. [1]
 (b) Identify structure **B**. [1]
 (c) Name **two** substances that pass from the placenta to the foetus. [2]
 (d) What is the function of the amniotic fluid? [1]

H 3 Figure 5.7 shows how the level of oestrogen changes during the menstrual cycle.

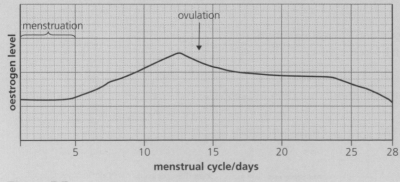

Figure 5.7

Exam practice answers at **www.hoddereducation.co.uk/myrevisionnotesdownloads**

(H)

(a) (i) What is the evidence that oestrogen stimulates ovulation? [1]

   (ii) Name **one** other effect of oestrogen. [1]

(b) (i) Name **one** other hormone involved in controlling the menstrual cycle. [1]

   (ii) Give **one** function of this hormone. [1]

4 (a) Copy and complete Table 5.2, showing some of the advantages and disadvantages of methods of contraception. [2]

Method of contraception	Advantage	Disadvantage
Condom	Prevents STIs	
Male sterilisation (vasectomy)	(Virtually) 100% reliable	

(b) (i) In terms of how they work, give **one** similarity between the contraceptive pill and an implant. [1]

   (ii) Give **one** advantage of the implant over the contraceptive pill. [1]

(c) (i) Explain the process of female sterilisation. [2]

   (ii) Explain how female sterilisation prevents pregnancy. [1]

Answers online

ONLINE ☐

6 Variation and adaptation

Living organisms that belong to the same **species** (type) resemble each other, but usually differ in a number of ways. These differences are called **variation**.

Variation

REVISED

Variation can be:
- **genetic** – due to our genes
- **environmental** – due to the environment or lifestyle
- due to a combination of both. For example, you have genes for a particular height, but your actual height also depends on your health and diet.

Variation can be **continuous** or **discontinuous**, as shown in Table 6.1.

Table 6.1 Continuous and discontinuous variation

Variation	Description	Examples
Continuous	Gradual change in a feature with no clearly distinct groups; no clear boundaries	Height/length
Discontinuous	Individuals can be grouped into a small number of distinct groups easily with no overlap	Tongue rolling/ hand dominance

Discontinuous variation is usually **genetic**, for example eye colour and blood group. Continuous variation is often both **genetic and environmental**.

> **Continuous variation**: The type of variation in which there is a gradual change in a feature with no distinct groups.
>
> **Discontinuous variation**: The type of variation in which all the individuals can be clearly divided into a small number of groups with no overlap.

> **Exam tip**
>
> Genetics and the environment are the *causes* of variation. Continuous and discontinuous are the *types* of variation.

> **Exam tip**
>
> In exam questions continuous variation is often represented by a histogram and discontinuous variation by a bar chart.

> **Now test yourself** TESTED
>
> 1 Give the **two** main causes of variation in populations.
> 2 Name the **two** types of variation.
>
> **Answers on p. 150**

⊕ Natural selection

REVISED

All living organisms are **adapted** for living in their normal environment. For example, a polar bear is camouflaged against the white snow and ice and its thick fur protects it against the cold. If organisms were not adapted they couldn't survive.

However, within any one species some organisms are better adapted to survive than others. This is because the different organisms of a species vary from each other. The better adapted organisms are more likely to survive than the less well adapted ones, which makes them more likely to reproduce and pass their (advantageous) genes on to their offspring. This is **natural selection**.

Natural selection can be particularly obvious in species where there is **competition for resources**.

> **Natural selection**: The process in which the better adapted individuals survive (at the expense of the less well adapted ones) long enough to reproduce and pass on their genes to their offspring.

HOne of the best examples of natural selection is **antibiotic resistance in bacteria**. The resistant phenotypes are not killed by antibiotics and so survive but the non–resistant bacteria are killed by antibiotics. The resistant bacteria are then able to survive and pass their (resistant) genes on to the next generations (Figure 6.1).

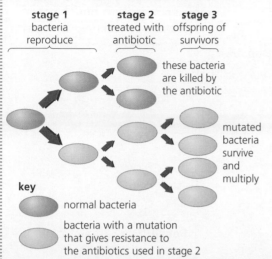

key

normal bacteria

bacteria with a mutation that gives resistance to the antibiotics used in stage 2

Figure 6.1 Antibiotic resistance in bacteria

In an exam, when provided with any example and suitable data, you need to be able to describe the process of natural selection.

Worked example

Pesticides are chemicals used to kill harmful insects. This example concerns mosquitoes.
- In many insect species some of the insects (<10%) are resistant to (and so not harmed by) pesticides because of a mutation.
- Most insects are not resistant and would be killed if they were sprayed by a pesticide.
- If a swarm of mosquitoes was sprayed with a pesticide, the resistant ones would survive and all the non-resistant ones would be killed (Figure 6.2).

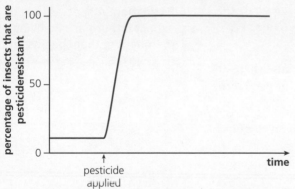

Figure 6.2 Pesticide resistance in mosquitoes

Using this example, explain what is meant by:
(a) variation within phenotypes
(b) differential survival
(c) natural selection.

Answer

(a) Some mosquitoes are pesticide resistant and some are not.
(b) The pesticide-resistant mosquitoes survive but the non-resistant ones are killed.
(c) Only the best-adapted mosquitoes (the pesticide-resistant ones) survive, reproduce and pass their genes on to the next generation.

The link between natural selection and evolution

Charles Darwin used his theory of natural selection to explain the process of **evolution**.

- Natural selection can explain how species have changed gradually over time in a process called evolution.
- This happens because certain features in the species are favoured.
- Eventually the species may be very different from how it started out.
- Evolution is a continuing process; natural selection is always happening and all species change very gradually over a long time period.
- Evolution can also result in **new species** developing.

There are a number of reasons why not everyone accepts the theory of evolution. These include the idea that:

- it contradicts some religious beliefs
- the very long time scales involved mean that it is difficult to see evolution actually happening.

> **Evolution**: The continuing process of natural selection that leads to the change in a species over time (or the formation of a new species).

Fossils

Fossils are the remains of living organisms that have been preserved (usually in rocks) for millions of years.

Fossils not only provide evidence for evolution, they can show the different changes that took place in a species over time (i.e. they can show how evolution occurred).

> **Fossil**: The remains of a living organism that has been preserved (usually in rock) for millions of years.

> ## Now test yourself
> TESTED
>
> 3 Define the term 'evolution'.
> 4 State **two** ways in which fossils can provide evidence for evolution.
>
> **Answers on p. 150**

> **Exam tip**
>
> As it is possible to date rocks to when they were formed (and when the organism was fossilised), it is possible to date the age of the fossil fairly accurately.

Extinction

Extinction means that there are no living examples of a species left. Extinct animals include the woolly mammoth and the dinosaurs. We know that extinct species once existed because of **fossils**.

An **endangered species** is one at **risk of extinction** because there are so few left. Examples of endangered species include the giant panda, the rhinoceros and some species of large cat.

Although extinction can happen naturally, humans have been responsible for the extinction of thousands of species and have put many more at risk. However, humans also have the potential to save endangered species by stopping them becoming extinct.

> **Extinction**: A species is extinct if there are no living members of that species left.

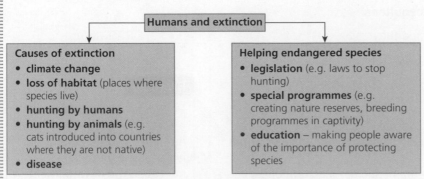

Humans and extinction

Causes of extinction
- **climate change**
- **loss of habitat** (places where species live)
- **hunting by humans**
- **hunting by animals** (e.g. cats introduced into countries where they are not native)
- **disease**

Helping endangered species
- **legislation** (e.g. laws to stop hunting)
- **special programmes** (e.g. creating nature reserves, breeding programmes in captivity)
- **education** – making people aware of the importance of protecting species

Figure 6.3 Our role in causing and preventing extinction

> **Exam tip**
>
> Humans cause environmental change, for example by adding predators to new environments, and this can lead to the extinction of some species.

Exam practice questions

1 (a) Explain what is meant by the term 'discontinuous variation'. [1]

(b) Give **one** example of discontinuous variation. [1]

2 Over time many species of predator and their prey have become more agile and able to run faster.

Use your understanding of natural selection to explain why the prey animals have become faster over time. [3]

3 In a typical pasture there may be a few plants that are resistant to high levels of copper in the soil. In these typical conditions, normal grasses grow better than the copper-resistant variety. However, in areas where the soil is contaminated with copper, the copper-resistant variety may make up over 90 per cent of the plants present. Explain how the increase in copper-resistant plants in contaminated soil demonstrates natural selection. [3]

4 The giant panda is at risk of extinction. It feeds on bamboo and a major reason for the panda becoming endangered is the reduction in bamboo forests. Suggest **two** ways in which humans can help panda numbers to recover. [2]

Answers online

ONLINE

7 Disease and body defences

Types of microorganism

A **communicable disease** is a disease that can be passed from one organism (person) to another.

Bacteria, **viruses** and **fungi** are the cause of most communicable diseases.

Table 7.1 provides information on some communicable diseases.

> **Communicable disease**: A disease that can be passed from one organism (person) to another.

Table 7.1 Communicable diseases

Microbe/disease	Type	Spread	Control/prevention/treatment
HIV (which leads to AIDS)	Virus	• Exchange of body fluids during sex • Infected blood	• Using a condom will reduce risk of infection, as will drug addicts not sharing needles • Currently controlled by drugs
Colds and flu	Virus	• Airborne (droplet infection)	• Flu vaccination for targeted groups
Human papilloma virus (HPV)	Virus	• Sexual contact	• HPV vaccination given to 12–13-year-old girls to protect against developing cervical cancer
Salmonella food poisoning	Bacterium	• From contaminated food	• Always cooking food thoroughly and not mixing cooked and uncooked foods can control spread • Treatment by antibiotics
Tuberculosis	Bacterium	• Airborne (droplet infection)	• BCG vaccination • If contracted, treat with drugs including antibiotics
Chlamydia	Bacterium	• Sexual contact	• Using a condom will reduce risk of infection • Treatment by antibiotics
Athlete's foot	Fungus	• Contact	• Reduce infection risk by avoiding direct contact in areas where spores are likely to be present, e.g. wear 'flip flops' in changing rooms/swimming pools
Potato blight	Fungus	• Spores spread in the air from plant to plant, particularly in humid and warm conditions	• Crop rotation and spraying plants with a fungicide

> **Exam tip**
>
> Potato blight is a plant disease that affects the potato and similar plants. All the other communicable diseases in Table 7.1 affect humans.

> **Exam tip**
>
> Communicable diseases are also described as **infectious** diseases.

The body's defence mechanisms

These mechanisms involve both stopping harmful microorganisms gaining entry to the body and destroying them in the blood.

1 The first stage of defence is stopping microorganisms from entering the body (Table 7.2).

Exam practice answers at **www.hoddereducation.co.uk/myrevisionnotesdownloads**

Table 7.2 Stopping microorganisms entering the body

Skin	Barrier that stops microorganisms from entering the body
Mucous membranes	Thin membranes in the nose and respiratory system that trap and expel microbes
Clotting	Closes wounds quickly to form a barrier that stops microorganisms gaining entry (also prevents loss of blood)

2 The role of **white blood cells** is to destroy microorganisms that have entered the body. There are two main ways this happens:

(a) **Lymphocytes** are white blood cells that produce antibodies when microorganisms enter the blood.

  ○ Microorganisms have special 'marker' chemicals on their surface called **antigens**.
  ○ These antigens cause the lymphocytes (white blood cells) to produce **antibodies**.
  ○ The antibodies are complementary in shape (like a lock and key) to the antigens.
  ○ They latch on to the antigens (microorganisms), linking them together.
  ○ This **immobilises** (clumps) the microorganisms and they can then be destroyed (Figure 7.1).
  ○ After an infection, the body produces **memory lymphocytes** that remain in the body for a very long time. These can respond quickly and produce high levels of antibodies very rapidly if the body is infected again by the same microorganism.

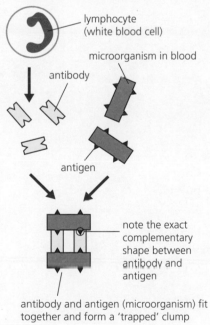

Figure 7.1 How antibodies work

(b) Once the microorganisms are clumped together, they are destroyed by a second type of white blood cell— the **phagocytes**. This process is called **phagocytosis** (Figure 7.2).

Exam tip

Clumping the harmful microorganisms (and then destroying them) prevents them from spreading around the body which will lead to reduced symptoms in the patient.

Lymphocyte: A type of white blood cell that produces antibodies.

Antigen: A distinctive marker on a microorganism that leads to the body producing specific antibodies.

Antibody: A structure produced by lymphocytes that has a complementary shape (and can attach) to antigens on a particular microorganism.

Memory lymphocyte: A special type of lymphocyte that can remain in the body for many years and produce antibodies quickly when required.

Exam tip

Because each type of microorganism has different types and **shapes** of antigen, each type of antibody has a unique shape that matches (is complementary to) the antigens. Therefore, there is a different type of antibody for each type of microorganism.

Phagocyte: A type of white blood cell that destroys microorganisms by engulfing them and then digesting them (phagocytosis).

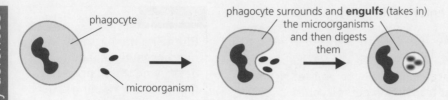

Figure 7.2 Phagocytosis

Now test yourself

TESTED

1 What is meant by the term 'communicable disease'?
2 State the type of microorganism that causes tuberculosis.
3 Name the type of white blood cell that produces antibodies.

Answers on p. 150

Active and passive immunity

Immunity is freedom from disease.

Active immunity is achieved by the body making its own antibodies (Figure 7.3).

> **Active immunity:** The type of immunity produced by the body making antibodies.

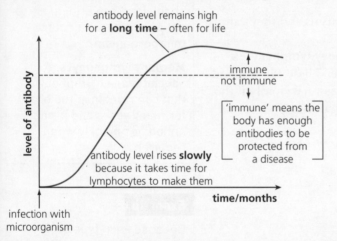

Figure 7.3 Active immunity through recovery from a disease

The pattern of immunity shown in Figure 7.3 is a typical response to being infected by a bacterium or a virus. The infected individual is often ill for a few days before the antibody numbers are high enough to provide immunity. This is known as the **primary response**.

However, if reinfection occurs, the memory lymphocytes can produce **large numbers** of antibodies **very quickly**. This is the **secondary response** (Figure 7.4).

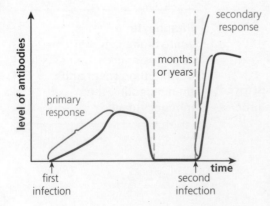

Figure 7.4 Primary and secondary responses to infection

Exam practice answers at **www.hoddereducation.co.uk/myrevisionnotesdownloads**

Vaccinations can also provide active immunity. Instead of getting the disease we are injected (vaccinated) with dead or weakened disease-causing microorganisms.

Worked example

(a) Why do the microorganisms in the vaccination need to be weakened or dead?

(b) Explain how a vaccination can provide immunity.

Answer

(a) If they were not, you would get the disease.

(b) The vaccination contains antigens (on the disease-causing microorganisms); these cause memory lymphocytes to be produced. If a subsequent infection occurs, the memory lymphocytes produce antibodies quickly and in high numbers. The antibodies latch on to the antigens (microorganisms) because they are complementary in shape. This immobilises the microorganisms. They are then destroyed by phagocytosis and the numbers of microorganisms do not get high enough to cause illness.

In **passive immunity** (Figure 7.5):

- we are given ready-made antibodies by injection
- the antibodies are not produced by the patient's body.

> **Passive immunity**: The type of immunity produced by injecting antibodies.

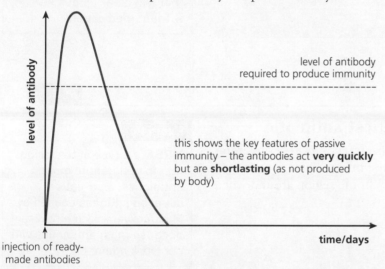

injection of ready-made antibodies

Figure 7.5 Passive immunity and the injection of ready-made antibodies

Passive immunity is **faster acting** than active immunity but the immunity does **not last as long**.

Passive immunity is an important option for people who catch a disease that could make them very ill before they get a chance to build up their own antibodies.

Antibiotics

REVISED

Antibiotics, such as penicillin, are chemicals **produced by fungi** that are used against bacterial diseases to **kill bacteria** or **reduce their growth**.

> **Antibiotic**: A chemical produced by fungi that kills bacteria.

Now test yourself

TESTED

4 Give **two** features of passive immunity.

5 What is an antibiotic?

Answers on p. 150

Exam tip

Antibiotics can kill bacteria but they have no effect on viruses.

ⒽAntibiotic resistance

Sometimes bacteria can evolve (change) so that antibiotics no longer have an effect. We say they have developed **antibiotic resistance**.

● The bacteria can **mutate**.
● Their DNA changes and the bacteria develop new properties.
● This change can make them **resistant** to antibiotics.
● Antibiotics will not work against these particular bacteria or cure diseases caused by them.

Antibiotic–resistant bacteria have been a particular problem in hospitals.

> **Antibiotic resistance**:
> Antibiotic-resistant bacteria cannot be killed by at least one type of antibiotic.

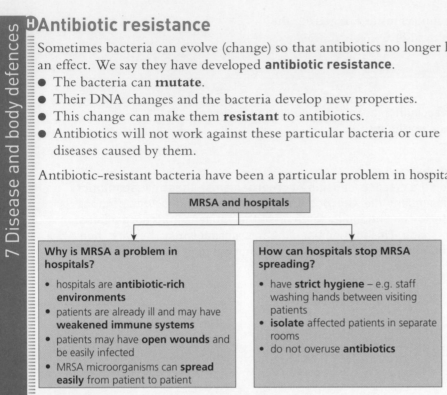

MRSA and hospitals

Why is MRSA a problem in hospitals?	**How can hospitals stop MRSA spreading?**
• hospitals are **antibiotic-rich environments** • patients are already ill and may have **weakened immune systems** • patients may have **open wounds** and be easily infected • MRSA microorganisms can **spread easily** from patient to patient	• have **strict hygiene** – e.g. staff washing hands between visiting patients • **isolate** affected patients in separate rooms • do not overuse **antibiotics**

Figure 7.6 MRSA (superbugs) and hospitals

> **Exam tip**
>
> Overuse of antibiotics has been a major factor in the development of bacterial resistance to antibiotics (and the development of 'superbugs'–Figure 7.6). We can reduce the use of antibiotics by not using them on viral diseases or non-serious bacterial infections that will clear up quickly without medication.

Development of medicines

REVISED

The discovery of penicillin: the first antibiotic

In 1928 **Alexander Fleming** was growing bacteria on plates containing a nutrient jelly (agar). One of his plates became infected by a fungus. Fleming noticed that bacteria could not grow in the region around where the fungus was growing (Figure 7.7).

The conclusion was that something (a chemical) was spreading from the fungus and killing the bacteria.

> **Exam tip**
>
> **MRSA** is a type of bacterium that is resistant to most antibiotics; a so-called superbug. Illness caused by MRSA is very difficult to treat because most antibiotics will not work against it.

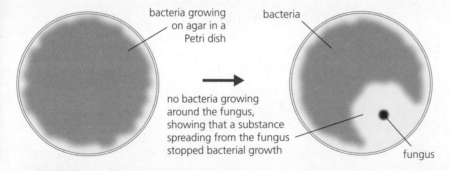

Figure 7.7 Fleming's discovery

Fleming was unable to isolate the chemical responsible but some years later two other scientists (**Florey** and **Chain**) were able to do so. This chemical was then commercially produced, as penicillin, and it became the first antibiotic.

Making new medicines and drugs

Medicines and medical drugs have to pass through a number of stages before they can be prescribed to the public. These stages are summarised in Table 7.3.

Table 7.3 The stages involved in testing medicines and drugs

Stage	What happens	Comment
In vitro testing	Testing on living cells in a laboratory	• Tests that the drug will combat disease and that the drug does not harm the cell • Allows testing before use on living organisms • An initial 'trial-and-error' process • Expensive because it needs highly trained scientists and involves expensive equipment
Animal testing	Testing on animals to check how it works on whole organisms	**Benefits** • Avoids testing on humans at this stage • Can check for **side effects** on living organisms **Disadvantages** • Animals are different from humans, so the drug may react differently when used in humans • Raises ethical issues
Clinical trials	Testing on small numbers of volunteers and then patients	Testing for side effects and how effective the drug is at doing what it is meant to do, and finding the best dosage
Licensing	The drug is licensed (given governmental approval)	It can be prescribed to treat members of the public

Exam tip

Drug development is **very expensive**. Table 7.3 states why *in vitro* testing is expensive, but the other stages are expensive too. Costs include the expense of keeping animals and payments to volunteers.

In vitro **testing**: The initial stage of testing drugs and medicines. It involves testing the drugs on cells and tissues in the laboratory.

Now test yourself

TESTED ☐

6 Name the **three** scientists involved in discovering and developing penicillin.
7 Place the **four** stages of drug and medicine development in order.

Answers on p. 150

Exam tip

You should remember the order in which drug development takes place: *in vitro*; animal testing; clinical trials; licensing.

Alcohol and tobacco

REVISED ☐

The misuse of alcohol and tobacco can harm health.

Misuse of drugs: alcohol

Many people can drink small amounts of alcohol without any harmful effects. However, drinking too much (especially binge drinking) can harm health.

Binge drinking is drinking too much alcohol on any one occasion or over a short period of time.

Drinking too much alcohol can:
• damage the **liver**
• cause damage to the brain and other organs in a developing **foetus** (if drinking too much during pregnancy).

Exam tip

As humans are a different species from the animals that a drug is tested on, there is no guarantee that a drug will work on humans until it is actually tested on them.

Side effect: An unwanted or unplanned effect of a drug on a person.

Binge drinking: Drinking a lot of alcohol over a short period of time.

Misuse of drugs: tobacco smoke

Smoking can seriously damage health, as summarised in Table 7.4.

Table 7.4 Harmful effects of tobacco smoke

Substance in cigarette smoke	Harmful effect(s)
Tar	Causes bronchitis (narrowing of airways in lungs), emphysema and lung cancer
Nicotine	Addictive and affects the heart rate
Carbon monoxide	Combines with red blood cells to reduce the oxygen-carrying capacity of the blood

Exam tip

You should be able to suggest ways in which people can reduce the harm caused by alcohol, e.g. drink less, switch to low-alcohol drinks, understand about units of alcohol, drink on fewer occasions.

Exam practice questions

1 (a) (i) Name the type of microorganism that causes colds. [1]
    (ii) Why do doctors not use antibiotics to treat colds? [1]
    (iii) Explain fully why many students in a school can become infected with a cold at the same time. [2]
  (b) (i) Name **one** type of disease in humans that is caused by a fungus. [1]
    (ii) Name **one** type of disease in plants that is caused by a fungus. [1]
2 (a) Describe what is meant by 'immunity'. [1]
  (b) Give **one** similarity and **one** difference between active and passive immunity. [2]
3 (a) (i) Describe what is meant by '*in vitro* testing'. [1]
    (ii) Suggest why it is so expensive. [1]
  (b) Apart from ethical considerations, give **one** argument for and **one** argument against animal testing. [2]
  (c) (i) What is meant by the term 'side effect'? [1]
    (ii) Suggest why a drug could be licensed for use, even though it has some side effects. [1]
4 John would like to give up smoking but finds it very hard to stop.
  (a) Name the chemical in cigarette smoke that makes it difficult for John to stop. [1]
  (b) Explain how the carbon monoxide in cigarette smoke can lead to smokers having a shortage of energy. [3]
Ⓗ 5 (a) (i) Describe what is meant by the term 'superbug'. [1]
    (ii) Give **one** example of a superbug. [1]
    (iii) Give the main reason for the development of superbugs. [1]
  (b) (i) Give **one** reason why superbugs are a particular problem in hospitals. [1]
    (ii) Give **two** ways in which hospitals are trying to reduce their incidence. [2]

Answers online

ONLINE

Exam practice answers at **www.hoddereducation.co.uk/myrevisionnotesdownloads**

8 Ecological relationships

Photosynthesis

REVISED

In **photosynthesis** plants make food using **light energy**. The light is trapped by **chlorophyll** in **chloroplasts** in the leaves of plants.

The word equation for photosynthesis is:

carbon dioxide + water → glucose + oxygen

H The balanced symbol equation is:

$$6CO_2 + 6H_2O \rightarrow C_6H_{12}O_6 + 6O_2$$

As photosynthesis requires (light) energy to work, it is an **endothermic reaction**.

The glucose produced during photosynthesis is usually converted into **starch** for storage. One way of showing that photosynthesis has taken place is by showing that starch is present in a leaf. This can be done using a **starch test**, as described in Table 8.1.

> **Photosynthesis**: A process in plants in which light energy is trapped by chlorophyll to produce sugars and starch (food).
>
> **Endothermic reaction**: A reaction that requires energy to be absorbed (taken in) to work.
>
> **Starch test**: A test to show whether or not starch is present in a plant leaf.

Table 8.1 Carrying out the test for starch

Step	Method	Reason
1	Put the leaf in boiling water	Kills the leaf and stops further reactions
2	Boil the leaf in alcohol (this must be done in a water bath because alcohol is flammable; use water from an electric kettle rather than using a Bunsen)	Removes chlorophyll (green colour) from the leaf; this makes any colour change when testing with iodine easier to see
3	Dip the leaf in boiling water again	Makes the leaf soft and less brittle (boiling in alcohol makes a leaf rigid)
4	Spread the leaf on a white tile and add iodine	If starch is present, the iodine will turn from yellow-brown to blue-black

Photosynthesis investigations

Before carrying out investigations into photosynthesis, it is usually necessary to **destarch** the plant. This involves leaving the plant in **darkness** (such as a dark cupboard) for **48 hours**. This is necessary to make sure that any starch found is starch that has been produced during the investigation.

> **Exam tip**
>
> If plants were not destarched in these investigations, the investigations would not be **valid** as it would be impossible to say whether any starch present was produced during the investigations or was there before they started.

Prescribed practical B3

Investigate the need for light and chlorophyll in photosynthesis by testing a leaf for starch

(a) Showing that light is needed for photosynthesis

- Destarch a plant.
- Partially cover a leaf on a plant with foil.
- Put the plant in bright light for at least 6 hours.
- Test the leaf for starch.

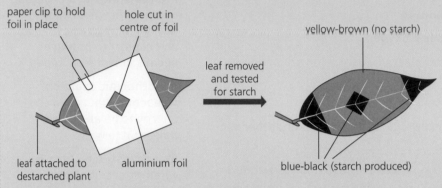

Figure 8.1 Showing that light is needed for photosynthesis

Figure 8.1 shows that starch is only produced in those parts of the leaf that receive light. The conclusion is that light is needed for photosynthesis.

(b) Showing that chlorophyll is needed for photosynthesis

Some plants have leaves that are part green and part white. These leaves are described as **variegated** leaves.

- Destarch a variegated plant.
- Put the plant in bright light for at least 6 hours.
- Test the leaf for starch.

The starch test will show that starch is only produced in those parts of the leaf that had chlorophyll (were green).

Showing that oxygen is produced during photosynthesis

Using apparatus similar to that shown in Figure 8.2 it is possible to demonstrate that oxygen is produced in photosynthesis.

The rate of photosynthesis will affect the rate at which the bubbles of oxygen are given off and this can be used to compare photosynthesis rates in different conditions.

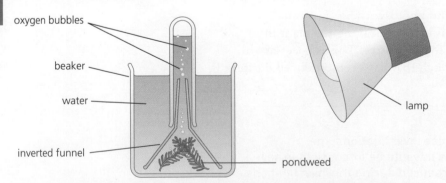

Figure 8.2 Showing that oxygen is produced during photosynthesis

Now test yourself

1 Write the word equation for photosynthesis.
2 What is a variegated leaf?

Answers on p. 150

TESTED

Exam practice answers at **www.hoddereducation.co.uk/myrevisionnotesdownloads**

Food chains and food webs

REVISED

A **food chain** describes the order in which energy passes through living organisms. In other words, it is a feeding sequence.

The **Sun** is the initial **source of energy** for all food chains.

A food chain always follows the order:

producer → primary consumer → secondary consumer → tertiary consumer

Table 8.2 Food chain terms

	Producer	Primary consumer	Secondary consumer	Tertiary consumer
Description	A plant that makes food by photosynthesis	An animal that feeds on a plant	An animal that feeds on a primary consumer	An animal that feeds on a secondary consumer
Example	Grass	Leaf-eating insect	Beetle	Insect-eating bird

The **arrows** in a food chain show the direction of the **transfer of energy** through a food chain as well as showing what eats what.

Food webs show how a number of food chains are interlinked. They are more realistic because very few consumers feed on only one type of plant or animal.

> **Food chain:** The order in which energy passes through a sequence of living organisms.
>
> **Producer:** A plant that makes food by photosynthesis. Producers always make up the first stage in a food chain.
>
> **Primary consumer:** An animal that eats plants.
>
> **Secondary (or tertiary) consumer:** An animal that eats other animals.
>
> **Food web:** A number of interlinked food chains.

Competition

REVISED

Living things compete with each other for resources.

For example, **plants** compete for:
- water
- light
- space to grow
- minerals.

Animals compete for:
- water
- food
- territory (space to live)
- mates.

> **Exam tip**
>
> Animals need plants both as an initial **food source** and also to provide **oxygen** (for respiration) through photosynthesis.

Competition among plant seedlings

In general, the more seedlings that are planted in a pot, the smaller the average mass of each seedling (Figure 8.3). This is because the seedlings are competing for light, space, minerals and water.

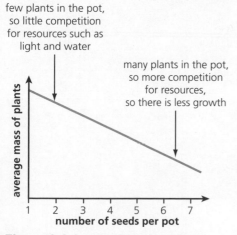

few plants in the pot, so little competition for resources such as light and water

many plants in the pot, so more competition for resources, so there is less growth

Figure 8.3 Competition in plants

> **Exam tip**
>
> In questions comparing the effect of seedling numbers in a pot and seedling growth, you are often asked about the factors that need to be kept constant to give valid results. These include using the same size of pot, using the same volume and type of compost and giving the pots the same environmental conditions (in other words, the same amount of water, the same level of light and keeping them at the same temperature).

Worked example

Jenna carried out an investigation to check whether the number of plants in a pot affected the growth of the plants. Table 8.3 shows some of the results.

Table 8.3 **Investigating the effect of planting density on plant mass**

Number of plants in the pot	Average mass of each plant/g	Total mass of all the plants in each pot/g
5	20	100
10	17	170
15	13	195
20	8	160
25	5	

(a) Copy and complete the table by determining the missing value.
(b) Describe and explain the effect of increasing the number of plants in the pot on the average mass of each plant.
(c) Which planting density should Jenna use if she wants to grow a large crop to help feed her pet rabbit?
(d) Jenna decided to repeat the investigation to get a second set of results. Explain the advantage of doing this.

Answer

(a) 25 × 5 = 125
(b) As the number of plants in the pot increases, the average mass of each plant decreases.
(c) 15 plants per pot; this gives the highest total mass.
(d) Increases reliability

Competitive invasive species

These are species that can compete so well that they can damage habitats and other species.

Competitive invasive species:
- are almost always **introduced to a country by humans** (they are 'foreign' organisms)
- **spread rapidly** when introduced into a region
- **outcompete** similar native species, usually causing them harm.

Two examples are:
- **grey squirrels** – introduced from America; they have spread rapidly, causing a decrease in the number of native red squirrels
- **rhododendron** – since being introduced this plant has spread rapidly; dense, evergreen leaves stop other species from growing underneath it.

> **Competitive invasive species:** A non-native species (introduced by man) that spreads rapidly, outcompeting native species.

Now test yourself

TESTED ☐

3 In a food chain, what name is given to the plants?
4 Give **two** things that animals compete for.
5 Define the term 'competitive invasive species'.

Answers on p. 151

Exam practice answers at **www.hoddereducation.co.uk/myrevisionnotesdownloads**

Monitoring change in the environment

REVISED

When monitoring the environment, we can use living (**biotic**) or non-living (**abiotic**) factors.

Abiotic (non-living) factors are important for monitoring global warming and also pollution.

Abiotic factors include:
- carbon dioxide levels
- size of ice fields
- water levels
- temperature
- pH.

Biotic (living) factors are also important. **Lichens** are sensitive plants that can be used to monitor pollution. Lichens will grow only in areas where pollution levels are low; they cannot grow in areas that have high levels of pollution.

> **Biotic factor**: A living organism, such as lichen, that can be used to monitor environmental change.
>
> **Abiotic factor**: A non-living factor, such as CO_2 level, that can be used to monitor environmental change.

Human activity on Earth

REVISED

Humans have been responsible for the fall in **biodiversity** on the Earth. Biodiversity is the term used to describe the **range of species** (different types of living organisms) in a particular area.

However, carefully managed human activity can have a positive effect on biodiversity.

Many of these positive approaches are described in Table 8.4.

> **Biodiversity**: The range/number of species in an area

Table 8.4 Human activity that can help increase biodiversity

Human activity	Example and explanation
Agriculture	• Replanting hedgerows – provides habitats for plant and animal species • Wide field margins – this also provides habitats for plant and animal species • Efficient use of fertilisers – prevents overuse as excess fertiliser can drain into waterways and cause pollution
Land use and management	• Reclaiming industrial sites – these can be reused for housing or industry and prevents building on 'greenfield' sites • Using 'brownfield' building sites – again, using land previously used for housing rather than building on greenfield sites • Planting sustainable woodlands – the planting of new woodlands can provide our needs for wood and also prevents deforestation of native woodland
Protecting fish stocks	Quotas, sanctuaries, limiting fishing to so many days per year, decommissioning some boats, limits on net sizes and large mesh sizes all help reduce numbers of fish caught so that stocks are conserved
Nature reserves	These protect rare species and rare habitats; can help educate the public about nature
International treaties	• Kyoto Protocol 1997 – limited success, as many countries didn't sign up to the agreement • Paris Agreement 2015 – plans to limit global warming to 2°C compared with pre-industrial levels; many more countries have signed up and it is legally binding

Now test yourself

TESTED

6 What is an abiotic factor?
7 What is meant by the term 'biodiversity'?
8 How does replanting hedgerows improve biodiversity?

Answers on p. 151

Exam practice questions

1 (a) (i) How would you destarch a plant? [1]
   (ii) Why is it necessary to destarch plants in a photosynthesis investigation? [1]
 (b) (i) In a starch test, give **one** safety precaution. [1]
   (ii) Explain fully why leaves are boiled in ethanol. [2]
2 Figure 8.4 shows a grassland food web.

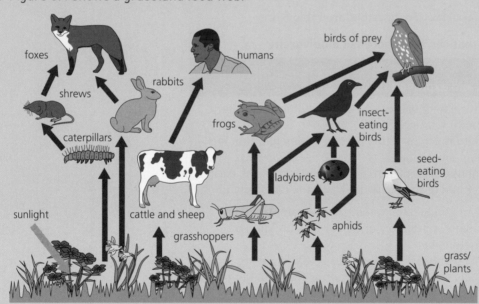

Figure 8.4

(a) What is the original source of energy for a food web? [1]
(b) Draw a food chain from the grassland food web in the diagram which involves three different organisms. [2]
(c) How many food chains are there involving four different organisms? [1]
(d) Suggest what would happen to the number of grasshoppers if the number of birds of prey decreased. Explain your answer. [2]

3 Figure 8.5 shows the relationship between the number of lichen plants on beech trees and their distance from a town in 1960 and 2010. During the investigation the numbers of lichen plants on tree trunks between the heights of 1 and 3 metres were counted. This was done for five trees at 5 km distances from the town.

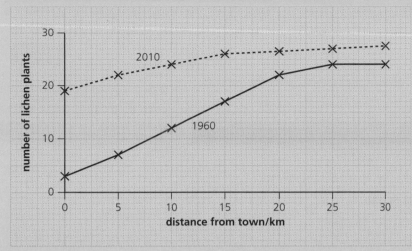

Figure 8.5

(a) State **two** trends shown by the graph.   [2]

(b) Suggest an explanation for **one** of the trends.   [1]

(c) Use the information provided to state **two** things that were done to ensure that the results were valid.   [2]

(d) State **one** thing that could be done to increase the reliability of the results.   [1]

4 Nature reserves help protect biodiversity.

(a) What is meant by biodiversity?   [1]

(b) For an area to be selected as a nature reserve, it usually has a number of features. Suggest **one** feature that a nature reserve should have.   [1]

(c) Describe and explain the effect that rhododendron will have on the biodiversity of a woodland floor.   [3]

Answers online

ONLINE ☐

9 Acids, bases and salts

Hazard symbols

REVISED

Table 9.1 shows the main **hazard symbols** and explains what they mean.

Table 9.1 Hazard symbols

Hazard	Explanation
Corrosive	Can burn the skin
Toxic	Can cause harm by poisoning
Explosive	Can explode
Flammable	Can catch fire easily
Caution	Must be handled with care

> **Hazard symbol**: A symbol used on a chemical to warn of danger.

corrosive   toxic   explosive   flammable   caution

Figure 9.1 Hazard symbols

Hazard symbols are used because:
- they have **greater visual impact** than words
- they are **easily recognised**
- they can be understood irrespective of **language**.

> **Exam tip**
>
> If asked to name a hazard represented by the toxic symbol, you must state 'toxic'; 'poisoning' does not usually gain credit because this is its effect.

Acids and bases

REVISED

A substance with pH 7 is described as having a **neutral pH**. **Acids** have a pH **lower than 7** and **bases** have a pH **higher than 7**. A base that dissolves in water is also called an **alkali**.

Both acids and alkalis can be strong or weak, as shown in Table 9.2.

Table 9.2 The pH of acids and alkalis

pH	0	1	2	3	4	5	6	7	8	9	10	11	12	13	14
Strength	Strong acid			Weak acid				Neutral	Weak alkali				Strong alkali		
Examples	Hydrochloric acid Sulfuric acid			Ethanoic acid					Ammonia				Sodium hydroxide Potassium hydroxide		

> **Acid**: A solution with a pH less than 7.
>
> **Base**: An insoluble alkali, with a pH more than 7.
>
> **Alkali**: A solution with a pH more than 7.

Acids and alkalis/bases can be **useful in the home**.

For example:
- vinegar is acidic – it contains ethanoic acid
- lemon juice is acidic – it contains citric acid
- baking soda is a base – it contains sodium hydrogencarbonate

> **Exam tip**
>
> The closer an acid is to pH 0, the stronger an acid it is; the closer an alkali is to pH 14, the stronger an alkali it is.

- *Milk of Magnesia* is an alkali – it contains the base magnesium hydroxide
- many cleaning products are alkalis – they contain ammonia
- many oven and drain cleaners are alkalis – they contain sodium hydroxide.

Indicators and pH

Indicators can tell us whether a substance is neutral, acid or alkaline. They usually do this by having different colours when in different solutions.

Natural indicators are those that occur naturally. Beetroot, blackcurrant and red cabbage plants contain dyes that change colour depending on whether they are in acidic, neutral or alkaline solutions.

Natural indicators can be made as follows:
1 Chop the plant or leaves into small pieces and put in a mortar.
2 Add a small amount of water.
3 Use a pestle to grind the mixture to release the dye.
4 The dye and the water form a solution that can be filtered into a small beaker.

This solution can act as an indicator because it turns different colours when added to neutral, acidic or alkaline solutions.

Universal indicator can do more than a natural indicator. It can tell us the **strength** of an alkali or an acid. It will turn a different colour in a weak acid (e.g. citric acid) than it will in a strong acid (e.g. hydrochloric acid).

Table 9.3 shows the colours of universal indicator over a range of pH values.

> **Indicator**: A chemical that can change colour to show whether a substance is an acid, neutral or an alkali.

Table 9.3 The colour of universal indicator at different pHs

pH	0	1	2	3	4	5	6	7	8	9	10	11	12	13	14
Universal indicator colour	Red			Orange		Yellow		Green	Green/ blue		Blue		Purple		
Strength	Strong acid			Weak acid				Neutral	Weak alkali				Strong alkali		

There are other ways of measuring the pH of a solution:
1 Litmus paper
   There are two types of litmus paper:
   - **blue** litmus paper turns **red** in **acid**
   - **red** litmus paper turns **blue** in **alkali**.
2 A pH sensor (pH probe)
   This is an electronic device that gives a numerical value (rather than a colour) for pH. Connecting the sensor to a data logger allows pH values to be transferred to a computer.
   A pH sensor has a number of **advantages**:
   - Numerical readings from pH sensors are more accurate than visually estimating colours and the pH value may be given to two decimal places.
   - A data logger can record continuous values; this is important when recording pH changes over a period of time.

Now test yourself

TESTED ☐

1 Give **one** example of a weak acid.
2 Give **one** example of a natural indicator.

Answers on p. 151

Neutralisation

When an alkali is added to an acid, the pH changes from a low value (the value of the acid) to a higher pH value.

The reaction is described as **neutralisation**. Figure 9.2 shows how to carry out a neutralisation reaction.

> **Neutralisation**: The reaction between an alkali (base) and an acid. Also refers to the process of using alkali to 'neutralise' the effect of acid, for example in indigestion remedies.

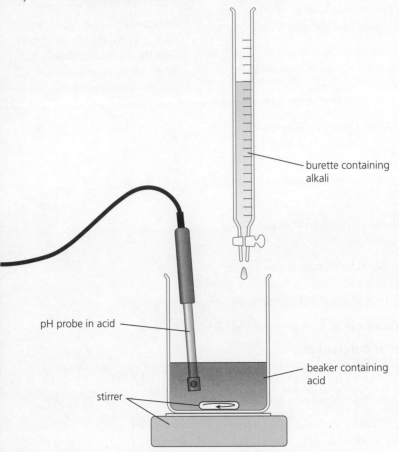

burette containing alkali

pH probe in acid

beaker containing acid

stirrer

> **Exam tip**
>
> A **burette** allows the amount of alkali added to be measured accurately.

> **Exam tip**
>
> The use of a pH probe (sensor) allows the pH value to be measured accurately at each stage of the reaction. This is better than trying to estimate colours at intervals using universal indicator.

Figure 9.2 A neutralisation reaction

The type of graph produced when adding an alkali (e.g. sodium hydroxide) to an acid (e.g. hydrochloric acid) is shown in Figure 9.3.

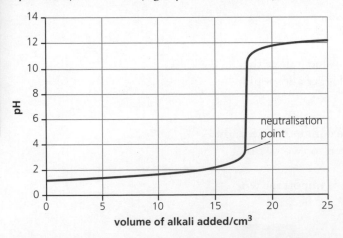

neutralisation point

Figure 9.3 Graph of pH changes in a neutralisation reaction

Prescribed practical C1

Carry out practical work to follow a neutralisation reaction by monitoring pH

Examples of neutralisation

Acid indigestion

The human stomach contains hydrochloric acid, a strong acid with a pH around 2. This acid kills microorganisms reaching the stomach in food and also helps the process of digestion. However, too much acid can cause **acid indigestion**. People can take substances that neutralise excess stomach acid and cure the indigestion.

Indigestion tablets contain weak bases such as oxides or hydroxides (such as magnesium hydroxide) and/or carbonates (such as sodium hydrogencarbonate).

For example, magnesium hydroxide is an antacid in *Milk of Magnesia*.

| magnesium hydroxide | + | hydrochloric acid | → | magnesium chloride | + | water |

Toothpaste

Bacteria in the mouth produce acid as a by-product of feeding on food. This acid can cause tooth decay. Many toothpastes contain a weak alkali to neutralise this acid.

Baking soda (bicarbonate of soda) is sodium hydrogencarbonate, which can be added to toothpaste for this purpose.

The chemistry of neutralisation

REVISED

Neutralisation is the reaction between an acid and an alkali (base), forming a **salt** and **water**. The general word equation for this reaction would be:

acid + base → salt + water

In an example of neutralisation described earlier:

| hydrochloric acid | + | magnesium hydroxide | → | magnesium chloride | + | water |
| (acid) | | (alkali) | | (salt) | | |

- The first part of the name of the salt is the same ('magnesium') as the metal in the alkali (base).
- The second part of the name of the salt is 'chloride' because it comes from the acid (hydrochloric acid).

> **Exam tip**
>
> The salt formed from a reaction with hydrochloric acid is a **chloride**. If sulfuric acid is used, a **sulfate** will be formed.

> **Worked example**
>
> Write the word equation for the neutralisation of sulfuric acid by magnesium hydroxide.
>
> *Answer*
>
> sulfuric acid + magnesium hydroxide → magnesium sulfate + water
>
>   (acid)   (alkali)   ·   (salt)

Using metal carbonates or hydrogencarbonates in neutralisation reactions

Earlier in the chapter we saw that sodium hydrogencarbonate (baking soda) can be used as an antacid to neutralise excess stomach acid. The word equation for this reaction is:

sodium hydrogencarbonate + hydrochloric acid → sodium chloride + carbon dioxide + water

This reaction neutralises the excess hydrochloric acid and the carbon dioxide gas escapes out of the body as a burp (wind).

> **Exam tip**
>
> When a **carbonate** or **hydrogencarbonate** is used in a neutralisation reaction, **carbon dioxide** is also produced (as well as the salt and water). As a general word equation
>
> acid + metal carbonate/hydrogencarbonate → salt + water + carbon dioxide

Other reactions with acids

An **acid** can react with a **metal** to give a **salt** and **hydrogen**:

acid + metal → salt + hydrogen

For example:

sulfuric acid + calcium → calcium sulfate + hydrogen

and

hydrochloric acid + magnesium → magnesium chloride + hydrogen

Reactions involving acids are summarised in Table 9.4.

Table 9.4 Reactions involving acids

Reactants	Products	Observations
Acid + base/alkali	Salt + water	Temperature rises
Acid + metal carbonate	Salt + water + carbon dioxide	Bubbles of carbon dioxide; temperature rises
Acid + metal hydrogencarbonate	Salt + water + carbon dioxide	Bubbles of carbon dioxide; temperature rises
Acid + metal	Salt + hydrogen	Bubbles of hydrogen; temperature rises

> **Now test yourself**
>
> 3 Name the type of reaction that produces a salt and water.
> 4 Name the products formed when an acid reacts with a metal.
>
> **Answers on p. 151**
>
> TESTED

Testing for gases

Depending on the reaction with an acid, carbon dioxide or hydrogen can be formed. It is possible to test for the presence of these gases (and other gases). Some of the most common lab tests for gases are described below.

Testing for carbon dioxide

Limewater is used to test for carbon dioxide gas. If carbon dioxide is **bubbled through limewater**, the **colourless** limewater will become **milky/cloudy** (Figure 9.4).

> **Limewater**: The chemical used to test for the presence of carbon dioxide

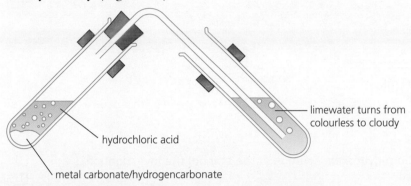

limewater turns from colourless to cloudy

hydrochloric acid

metal carbonate/hydrogencarbonate

Figure 9.4 Using limewater to test for carbon dioxide

> **Exam tip**
>
> Limewater is a test for carbon dioxide because carbon dioxide is the only gas that turns it from colourless to cloudy/milky.

Testing for hydrogen

Insert a **burning splint** into a test tube containing hydrogen and the hydrogen will ignite and burn rapidly with a '**squeaky pop**' sound.

Testing for oxygen

Insert a **glowing splint** (a splint in which the flame has just been extinguished) into a test tube containing oxygen. If oxygen is present, the splint will **relight** (ignite).

> **Exam tip**
>
> Carbon dioxide, hydrogen and oxygen are all colourless gases.

Exam practice questions

1 Table 9.5 provides information about acids and alkalis.
   (a) Copy and complete the table. [2]

Table 9.5

pH	0	1	2	3	4	5	6	7	8	9	10	11	12	13	14
Strength	Strong acid			_____ [1]				Neutral	Weak alkali				Strong alkali		
Examples	Hydrochloric acid Sulfuric acid			ethanoic acid					_____ [1]				Sodium hydroxide Potassium hydroxide		

   (b) What is meant by the term 'base'? [1]

→

2 The pH change when hydrochloric acid is added to sodium hydroxide is shown in Figure 9.5.

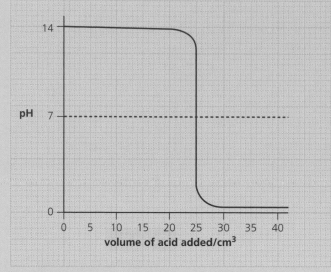

Figure 9.5

(a) What was the pH value of the sodium hydroxide solution at the start of the investigation? [1]
(b) Name the type of reaction involved. [1]
(c) Suggest why a pH sensor was used in this investigation. [1]
(d) Name a suitable piece of apparatus that could have been used to add the hydrochloric acid. [1]
3 (a) Describe what causes indigestion. [1]
(b) Explain how sodium hydrogencarbonate can cure indigestion. [3]
(c) Why can indigestion remedies cause people to suffer 'wind'? [1]
4 Sherbet is a fizzy powder containing a number of substances including an edible acid and a base. When taken, it reacts with water in the mouth to produce a fizzing sensation.
(a) Copy and complete the word equation for this reaction:

sodium hydrogencarbonate + citric acid → _____ **+** _____ **+** _____ [3]

(b) Suggest what causes the fizzing sensation. [1]
5 Hydrogen and carbon dioxide are both colourless gases.
(a) (i) Give **one** example of a reaction that produces hydrogen. [1]
(ii) Give **one** example of a reaction that produces carbon dioxide. [1]
(b) Complete Table 9.6 showing the tests for hydrogen and carbon dioxide.

Table 9.6

Gas	Test	Result if gas present
hydrogen	burning splint	_____[1]
carbon dioxide	_____[1]	turns from colourless to cloudy

[2]

Answers online

ONLINE

10 Elements, compounds and mixtures

Solids, liquids and gases

There are three states of matter: solids, liquids and gases (Figure 10.1).

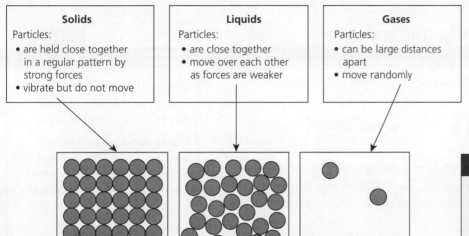

Solids

Particles:
- are held close together in a regular pattern by strong forces
- vibrate but do not move

Liquids

Particles:
- are close together
- move over each other as forces are weaker

Gases

Particles:
- can be large distances apart
- move randomly

Figure 10.1 States of matter

> **Exam tip**
>
> Solids have a fixed volume and shape; liquids have a fixed volume but not shape; gases have neither a fixed volume nor shape.

Changes of state

Changes of state are physical reactions. Examples include:
- **melting** – solids changing to liquids
- **boiling/evaporation** – liquids changing to gases
- **freezing** – liquids changing to solids
- **condensing** – gases changing to liquids
- **sublimation** – gases changing directly into solids or solids directly into gases.

The **melting point** is the temperature at which a solid changes into a liquid.

The **boiling point** is the temperature at which a liquid changes into a gas.

Figure 10.2 summarises the changes of state.

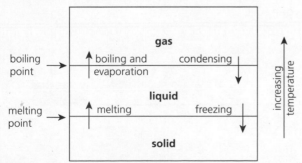

Figure 10.2 Changes of state

> **Melting**: The process during which a solid changes into a liquid.
>
> **Boiling (evaporation)**: The process during which a liquid changes into a gas.
>
> **Freezing**: The process during which a liquid changes into a solid.
>
> **Condensing**: The process during which a gas turns into a liquid.
>
> **Sublimation**: The process during which gases change directly into solids or solids change directly into gases.

> **Exam tip**
>
> **Sublimation** misses out the liquid state as gases change directly into solids or solids change directly into gases.

(H) Students should be able to classify substances as solids, liquids or gases given melting point and boiling point data.

Now test yourself

TESTED

1 Name the term used to describe a liquid changing into a solid.
2 Define the term 'condensation'.

Answers on p. 151

Elements, compounds and mixtures

REVISED

An **element** is a pure chemical substance that is made from only one type of **atom**. An element cannot be chemically broken down into anything simpler.

A **compound** is a substance that has two or more elements chemically joined together.

A **mixture** contains two or more different substances (elements or compounds or a combination of each). The different substances in a mixture can be separated using a variety of methods as they are not chemically joined.

Table 10.1 gives some examples of elements, compounds and mixtures.

Element: A substance which contains only one type of atom.

Compound: A substance that has the atoms of two or more different elements chemically bonded (joined) together.

Mixture: Two or more substances together that are not chemically bonded (joined).

Table 10.1 Examples of elements, compounds and mixtures

	Example	Formula	Description
Element	Hydrogen	H_2	Hydrogen is a gas formed of hydrogen atoms
	Oxygen	O_2	Oxygen is a gas in which the smallest particle is two oxygen atoms bonded together
Compound	Sodium chloride	$NaCl$	Sodium chloride is a salt formed from two elements, sodium and chlorine
	Sulfuric acid	H_2SO_4	The smallest unit in sulfuric acid contains two hydrogen atoms, one sulfur atom and four oxygen atoms
Mixture	Air		Air is a mixture containing a number of gases including nitrogen (N_2), oxygen (O_2), carbon dioxide (CO_2) and water vapour (H_2O)

Chemical formulae and equations

REVISED

Chemical formulae

You need to be familiar with, and be able to recognise, common chemical formulae. The formulae for some of the substances used earlier in the chemistry section are:
- $NaOH$, sodium hydroxide
- HCl, hydrochloric acid

- NaCl, sodium chloride
- H_2O, water.

You need to be able to work out:
- the number of **elements** in a chemical formula; for example, NaCl has two elements (Na (sodium) and Cl (chlorine)) while NaOH has three different elements (Na (sodium), O (oxygen) and H (hydrogen))
- the number of **atoms**; NaCl has one atom of sodium (Na) and one of chlorine (Cl), while water (H_2O) has two atoms of hydrogen and one of oxygen.

A more complex formula is $Mg(OH)_2$. Here there is one molecule of magnesium (Mg) but two each of oxygen (O) and hydrogen (H).

Atom: The smallest sub-unit of an element; the smallest part of an element that can exist

Now test yourself

TESTED

3 Define the term 'compound'.
4 How many atoms are represented by the formula HCl?

Answers on p. 151

Exam tip

When working out the numbers of atoms, you should remember that if the number is a subscript (small and low down) it refers to the element immediately before it.

Ⓗ Symbol equations using chemical formulae

The symbol equation for the neutralisation reaction between sodium hydroxide and hydrochloric acid is:

$$NaOH + HCl \rightarrow NaCl + H_2O$$

This equation is balanced because there are the same numbers of each type of atom on each side. There is one Na atom on each side, one Cl atom on each side and there are two H atoms on the left and two on the right (H_2).

When writing symbol equations it is often necessary to balance the sides; you need to make the numbers of atoms on each side match.

Worked examples

For the neutralisation reaction covered earlier:

hydrochloric acid + magnesium hydroxide → magnesium chloride + water

the balanced symbol equation is:

$$Mg(OH)_2 + 2HCl \rightarrow MgCl_2 + 2H_2O$$

The '2's before the HCl and before the H_2O are balancing numbers; they are needed to balance the numbers of atoms on each side. '2HCl' means that there are two hydrogen atoms and two chlorine atoms. In '2H_2O', there are four hydrogen atoms and two oxygen atoms.

The brackets around the 'OH' and the '$_2$' after it means that there are two oxygen atoms and two hydrogen atoms.

(a) Calcium carbonate reacts with hydrochloric acid to produce calcium chloride, carbon dioxide and water.
Balance the symbol equation for this reaction:

$$CaCO_3 + HCl \rightarrow CaCl_2 + CO_2 + H_2O$$

10 Elements, compounds and mixtures

CCEA GCSE Science Single Award   57

(b) Hydrogen burns in oxygen to produce water.
   Balance the equation

$$H_2 + O_2 \rightarrow H_2O:$$

Answer

(a) • There is one calcium (Ca) atom on each side – no balancing needed.
   • There is one carbon atom (C) on each side – no balancing needed.
   • There are three oxygen (O) atoms on each side – no balancing needed.
   • There is one hydrogen (H) atom on the left and two on the right (balancing needed).
   • There is one chlorine atom (Cl) on the left and two on the right (balancing needed).
   • The equation can be balanced by putting a '2' before the HCl on the left; remember this means there are two 'H' atoms and two 'Cl' atoms.

The balanced symbol equation is, therefore:

$$CaCO_3 + 2HCl \rightarrow CaCl_2 + CO_2 + H_2O$$

(b) $2H_2 + O_2 \rightarrow 2H_2O$

State symbols

Balanced symbol equations can contain the **state symbols** to show the state of each of the substances involved: solid (s), liquid (l), gas (g) and aqueous solution (aq).

For example:

$$CaCO_3(s) + 2HCl(aq) \rightarrow CaCl_2(aq) + H_2O(l) + CO_2(g)$$

> **Exam tip**
>
> **Liquid** refers to a pure substance in a liquid state; **aqueous** is the term used to describe a solvent (water) with a solute dissolved in it.

Separating mixtures

REVISED

Pure and impure substances

- **Pure** substances contain a single element or compound and are not mixed with any other substance.
- **Impure** substances contain a mixture of different substances.

Methods of separation

Some key terms are described in Table 10.2.

Table 10.2 Some key terms

Term	Description	Example
Soluble	A solid is soluble if it dissolves in a liquid	Glucose will dissolve in water
Insoluble	A solid that will not dissolve in a liquid	Iron filings will not dissolve in water
Solute	A solid that can dissolve in a liquid	Glucose
Solvent	The liquid that the solid dissolves in	Water
Solution	The solute and the solvent	Glucose solution

Exam practice answers at **www.hoddereducation.co.uk/myrevisionnotesdownloads**

Separation methods include:

- filtration
- crystallisation
- simple distillation
- paper chromatography.

Filtration

Filtration separates an **insoluble solid** from a **liquid**. For example, a mixture of sand and water can be separated by pouring it through a piece of filter paper in a filter funnel, as shown in Figure 10.3.

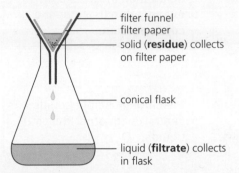

Figure 10.3 Filtration

Evaporation and crystallisation

This method is used to separate a **dissolved (soluble) solid** from the **liquid (solvent)** it is dissolved in (Figure 10.4). For example, salt (sodium chloride) can be recovered from a solution of salt in water by evaporating the liquid off.

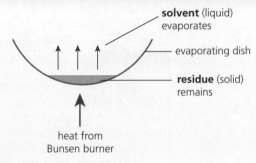

Figure 10.4 Evaporation

Crystallisation also separates a **dissolved solid** from its **solvent**. For example, it can be used to separate copper sulfate crystals from a solution of copper sulfate.

The mixture is heated in an evaporating dish, as above, to evaporate some of the solvent. This makes the solvent that remains more concentrated as it contains proportionally more solute. As the solvent is left to cool, not all the solute can remain dissolved, so some crystallises out to form crystals. The crystals can be separated from the rest of the solvent by filtration.

Simple distillation

Distillation is used to separate the **solvent** (liquid) from the **solution**, for example when separating pure water from sea water.

Figure 10.5 shows that distillation involves both evaporation and condensation.

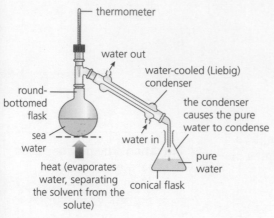

Figure 10.5 Simple distillation: separating pure water from sea water

H Liquids that mix together are called **miscible** liquids, such as water and alcohol; liquids that do not mix together are called **immiscible** liquids.

Simple distillation can also be used to separate two miscible liquids like alcohol and water as these two liquids have different **boiling points**. The liquid with the lower boiling point boils first and is collected following condensation.

Paper chromatography

Paper **chromatography** separates the different **soluble substances in a mixture**.

When chromatography paper (a chromatogram) is dipped into the solvent, the solvent moves up through the paper.

The different solutes in the mixture are carried by the solvent at different rates so they separate, as shown in Figure 10.6.

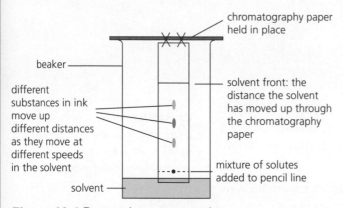

Figure 10.6 Paper chromatography

H The solvent is known as the **mobile phase** and the chromatogram (paper) is known as the **stationary phase**.

Different solutes will have different **R_f values**. The R_f value is the distance between the **leading edge of the spot** (solute) and the origin (point where the mixture was added to the paper) divided by the distance between the **solvent front** and the origin.

H

Worked example

Calculate the R_f value for spot (solute) B in Figure 10.7.

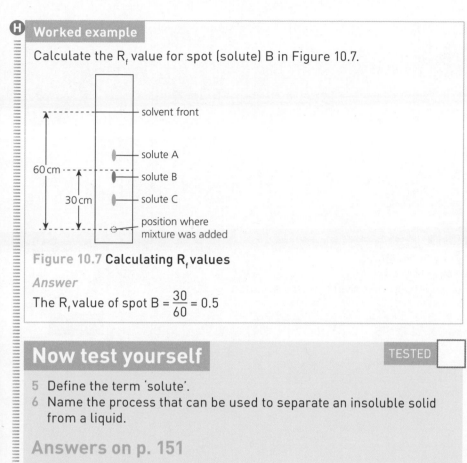

Figure 10.7 Calculating R_f values

Answer

The R_f value of spot B $= \dfrac{30}{60} = 0.5$

Now test yourself

TESTED ☐

5 Define the term 'solute'.
6 Name the process that can be used to separate an insoluble solid from a liquid.

Answers Answers on p. 151

Exam practice questions

1 (a) Copy and complete the sentences below.
The term _____ describes liquids changing to solids.
The melting point is the temperature at which a _____ changes into
a _____. [3]
(b) Explain what is meant by the term 'subliming'. [2]

2 NaNO$_3$ is sodium nitrate.
(a) How many elements are in sodium nitrate? [1]
(b) How many atoms does this formula represent? [1]

3 The word equation shows sodium hydroxide reacting with sulfuric acid:

sodium hydroxide + sulfuric acid → sodium sulfate + water

(a) Name the salt formed in this reaction. [1]
(b) Balance the symbol equation below: [2]

NaOH + H$_2$SO$_4$ → Na$_2$SO$_4$ + H$_2$O

4 (a) What is a pure substance? [1]
(b) Table 10.3 describes two important terms. Copy and complete the table. [2]

Term	Description
Soluble	A _____ [1] that can dissolve in a liquid
Solvent	_____ [1]

(c) Figure 10.8 shows how substances can be separated in chromatography.

→

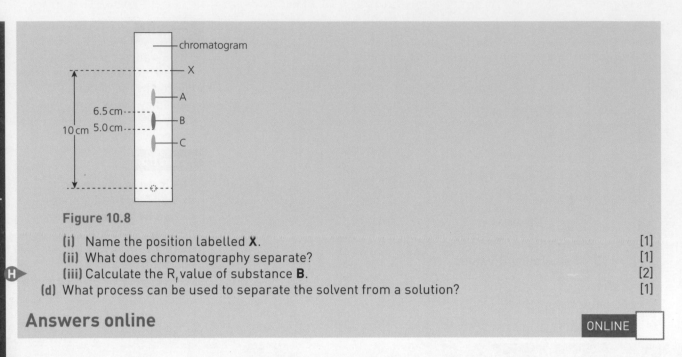

Figure 10.8

(i) Name the position labelled **X**. [1]

(ii) What does chromatography separate? [1]

(iii) Calculate the R$_f$ value of substance **B**. [2]

(d) What process can be used to separate the solvent from a solution? [1]

Answers online

ONLINE

11 Periodic Table, atomic structure and bonding

The structure of the atom

An **element** is a pure chemical substance that is made from only one type of **atom**.

Atoms are made up of three **subatomic** particles: **protons**, **electrons** and **neutrons**.

The protons and neutrons are found in a central **nucleus**. The electrons are arranged in **shells** that surround the nucleus (Figure 11.1).

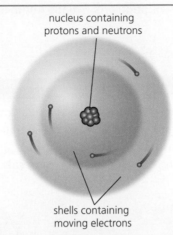

nucleus containing protons and neutrons

shells containing moving electrons

Figure 11.1 The structure of an atom

Atomic number and mass number

All atoms have an **atomic number**. This is the **number of protons** in the nucleus. This is the same as the number of electrons orbiting the nucleus.

All atoms also have a **mass number**. This is the total **number of protons and neutrons** in the nucleus.

The Periodic Table gives the atomic number and mass number of each element. For example, oxygen has an atomic number of 8 and a mass number of 16. Figure 11.2 shows how oxygen is represented in the Periodic Table.

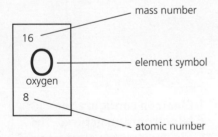

mass number

element symbol

atomic number

Figure 11.2 Oxygen atom description

Working out numbers of protons, electrons and neutrons in oxygen:
- atomic number = number of protons
- mass number = number of protons + number of neutrons

Oxygen atoms must have eight electrons because the number of electrons = the number of protons. They must also have eight neutrons because the number of neutrons = the mass number − the number of protons.

Atom: The smallest part of an element. Atoms are made up of even smaller sub-units called protons, neutrons and electrons.

Proton: The positive subatomic particle found in the nucleus of an atom.

Electron: The negative subatomic particle found in the electron shell(s) of an atom.

Neutron: An electrically neutral subatomic particle found in the nucleus of an atom.

Atomic number: The number of protons in the nucleus of an atom.

Mass number: The total number of protons and neutrons in the nucleus of an atom.

Exam tip

mass number = number of protons + number of neutrons

Worked example

An atom of potassium is represented as $^{39}_{19}K$.
(a) Calculate the number of electrons in an atom of potassium.
(b) Calculate the number of neutrons in an atom of potassium.

Answer

(a) 19 is the atomic number, which is the number of protons, so there must also be 19 electrons.
(b) 39 is the mass number and the number of neutrons = mass number − atomic number = 39 − 19 = 20.

Some properties of atoms

- **Mass**: Protons and neutrons have the same mass. However, the mass of an electron is negligible (relative to the mass of a proton and a neutron).
- **Charge**: Protons have a positive charge and electrons have a negative charge; neutrons have no charge. In atoms, the numbers of protons and electrons are equal. This means that an atom is **electrically neutral** and has **no charge**.

Table 11.1 summarises the properties of atoms.

Table11.1 Properties of atoms

Atomic particle	Relative mass	Relative charge
proton	1	+1
neutron	1	0
electron	(negligible)	−1

Exam tip

Remember that an atom has **no electrical charge** as the number of protons (with positive charge) is equal to the number of electrons (with negative charge).

Now test yourself

TESTED ☐

1 Name the three different types of subatomic particle found in an atom.
2 What is meant by the term 'mass number'?

Answers on p. 151

Electronic configuration (electronic structure)

The **electrons** in an atom orbit (or travel) around the nucleus in 'electron shells'. The **first shell** (the one closest to the nucleus) can have a maximum of **two electrons** in it. The **second** and **third shells** can hold a maximum of **eight electrons**. The shells fill up in order, so if an atom has six electrons, it will have two in the first shell and four in the second shell.

Figure 11.3 shows the electron arrangement (**electron configuration**) of a sodium atom to be 2, 8, 1. This means there are two electrons in the first shell, eight in the second and one in the third.

Electron configuration (electronic structure): The arrangement of electrons in the electron shells of an atom.

Electron shell: The zone(s) around the nucleus of an atom in which electrons are located.

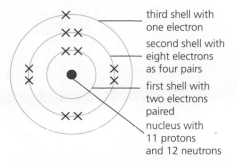

third shell with one electron

second shell with eight electrons as four pairs

first shell with two electrons paired

nucleus with 11 protons and 12 neutrons

Figure 11.3 The arrangement of electrons in a sodium atom

Exam tip

You need to be able to work out electron arrangements from atomic numbers. For sodium, the atomic number is 11, so there are 11 electrons in its atoms. Remember the shells fill up in turn. The first shell has the maximum two, the second shell the maximum eight and there is one electron in the third shell.

Elements 1–20

Figure 11.4 shows the electronic structure for atoms of elements with atomic numbers 1 to 20.

Hydrogen	Helium	Lithium	Beryllium	Boron
1	2	2, 1	2, 2	2, 3

Carbon	Nitrogen	Oxygen	Fluorine	Neon
2, 4	2, 5	2, 6	2, 7	2, 8

Sodium	Magnesium	Aluminium	Silicon	Phosphorus
2, 8, 1	2, 8, 2	2, 8, 3	2, 8, 4	2, 8, 5

Sulfur	Chlorine	Argon	Potassium	Calcium
2, 8, 6	2, 8, 7	2, 8, 8	2, 8, 8, 1	2, 8, 8, 2

Figure 11.4 The arrangement of electrons in atoms of elements 1–20. Note that hydrogen, with atomic number 1, has one electron; helium, with atomic number 2, has two electrons and so on until calcium, which has 20 electrons.

Following the rules of electron shells filling in order, calcium has an electronic structure of 2, 8, 8, 2 with the first three shells filled.

The Periodic Table

The **Periodic Table** lists all the known elements arranged in order based on their atomic number (Figure 11.5).

In the Periodic Table:
- the vertical columns are called **groups**; the elements in each group have similar properties
- the horizontal rows are called **periods**.

> **Periodic Table**: The table that lists the known elements in order of atomic number.

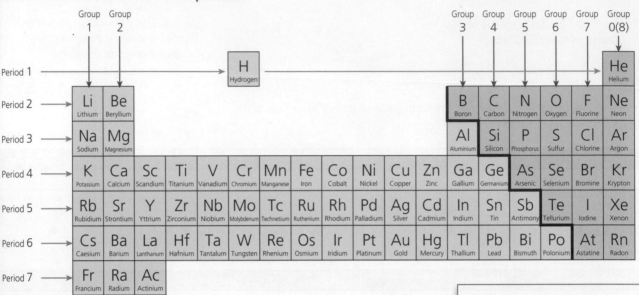

Figure 11.5 The Periodic Table

> **Group**: A vertical column in the Periodic Table. The group number gives the number of electrons in the outer shell of the elements in that group.
>
> **Period**: A horizontal row in the Periodic Table. The period number gives the number of electron shells in the atoms of the elements in that period.

Some other things you should know about the Periodic Table:
- The thick, black, stepped line on the right separates metals (on the left) from non-metals (on the right).
- The metallic character (properties) of the elements decreases as you move from left to right across the table.

Some of the groups have names as well as numbers:
- Group 1 is a group of reactive metals known as the **alkali metals**.
- Group 2 elements are the **alkaline earth metals**.
- Group 7 is a group of reactive non-metals called the **halogens**.
- Group 0 (or 8) contains the **noble gases**. These gases are chemically inert (non-reactive) and include helium, neon and argon. They are found in air but, because they are colourless and unreactive, they were discovered much later than many other elements.

> **Exam tip**
>
> When identifying a period number in the Periodic Table, it is important not to forget Period 1, the row at the top that just has hydrogen and helium in it.

Groups and periods

The **group (column) number** gives the number of **electrons** in the **outer shell** of the atoms. For example, sodium (Na) is in group 1. This means it has only one electron in the outer shell (electronic configuration: 2, 8, 1).

The **period (row) number** gives the number of **electron shells** in each atom. Sodium has three electron shells so is in period 3 (electronic configuration: 2, 8, 1).

The history of the Periodic Table

In 1869, the Russian chemist **Dmitri Mendeleev** developed a Periodic Table that is similar to the modern one.

- Mendeleev arranged the elements in **order of atomic mass** (or **atomic weight**).
- He left **gaps** for as-yet-**undiscovered elements**.
- He arranged the elements in **groups** (columns) and **periods** (rows).
- He **predicted the properties** of the yet-to-be discovered elements, e.g. the noble gases.

H The modern Periodic Table has updated Mendeleev's work by:

- arranging the elements in order of **atomic number**
- containing **more elements** than Mendeleev's table (all the gaps have been filled)
- adding the noble gases (group 0).

Compounds

REVISED

A **compound** is a substance that has two or more elements **chemically joined** together. Compounds are formed when two atoms share electrons or when electrons move from one atom to another.

When a compound is formed between a metal and a non-metal, the metal element (or the element to the left/lower down) keeps its name. The other element in the compound changes its name. For example, sodium and fluorine combine to form sodium fluor**ide**, NaF.

The same rule applies for simple compounds formed from non-metals. For example hydrogen (to the left of chlorine) and chlorine form hydrogen chlor**ide**.

As a general rule, substances with names ending in **-ide** have two elements and those ending in **-ate** have three elements, one of which is oxygen. For example, sodium fluor**ide** (NaF) is formed from sodium and fluorine; magnesium sulf**ate** ($MgSO_4$) is formed from magnesium, sulfur and oxygen. There is an exception to the –ide rule: hydroxides have three elements. Potassium hydroxide (KOH), for example, is formed from potassium, hydrogen and oxygen.

Some compounds, such as methane (CH_4), have only one name.

Some elements exist as molecules formed from two identical atoms (from the same element) joined together. Examples include oxygen (O_2) and hydrogen (H_2). They are known as **diatomic** molecules.

The alkali metals (group 1)

The **alkali metals** are so **reactive** that they need to be stored under oil to stop them reacting with air. They are soft, grey metals that can be cut easily with a knife.

> **Alkali metal**: A metal in group 1 of the Periodic Table

All the alkali metals react with water. If **lithium** is added to water it moves around vigorously on the surface (floats) and makes a hissing sound as **hydrogen gas** is produced. It dissolves and disappears as it is used up in the reaction. The word equation for lithium reacting with water is:

   lithium + water → lithium hydroxide + hydrogen

As you move down group 1 in the Periodic Table, the reaction of the metal with water becomes increasingly vigorous. This is because the metals are more reactive as you move down the table.

All the group 1 (alkali) metals react with water to form a metal hydroxide and release hydrogen gas. In effect:

   alkali metal + water → metal hydroxide + hydrogen

> **Exam tip**
>
> The alkali metals need to be stored in oil as they would react with moisture in air.

You need to know what happens in the following two reactions:

- **Sodium** and **water**: the sodium floats and moves about on water; bubbles of gas (hydrogen) are visible; the sodium may burn with a **yellow flame** before disappearing (being used up). This is a **vigorous** reaction.

   sodium + water → sodium hydroxide + hydrogen

> **Exam tip**
>
> Potassium reacts more vigorously with water than sodium as it is lower down the group in the Periodic Table.

- **Potassium** and **water**: the potassium floats and moves about on water; bubbles of gas (hydrogen) are visible; the potassium may burn with a **lilac flame** before disappearing (being used up). This is a **very vigorous** reaction.

   potassium + water → potassium hydroxide + hydrogen

H Remember that Higher-Tier candidates need to be able to write balanced symbol equations for the reactions covered in the specification, for example:

   $2Na + 2H_2O \rightarrow 2NaOH + H_2$

> **Exam tip**
>
> Remember when balancing equations that the number of atoms on each side for each element must be equal: 2Na, 4H and 2O.

The noble gases (group 0)

Group 0, the **noble gases**, are all **colourless**. Examples include helium and neon.

> **Noble gases**: The elements (gases) in group 0 of the Periodic Table

The noble gases are all very **unreactive**. They have a **full outer shell of electrons** which results in them being very stable. Elements without a full outer shell are more reactive as they are more likely to gain or lose electrons to give them a full outer shell.

Now test yourself

3 What name is used to describe an element in group 2 of the Periodic Table?
4 How many electrons do group 2 elements have in their outer shell?
5 Name the products formed when a group 1 element reacts with water.

Answers on p. 151

Exam practice answers at **www.hoddereducation.co.uk/myrevisionnotesdownloads**

Ionic and covalent bonding

When elements form compounds they can do this by **ionic** or **covalent** bonding.

Ionic bonding

When elements react to form compounds, their atoms become bonded together. In doing this, both elements gain **full outer shells of electrons** (also known as a noble gas electron structure).

For example, if an element in group 1 (with one electron in the outer shell) reacts with an element from group 7 (with seven electrons in the outer shell) an electron is transferred from the group 1 element to the group 7 element, resulting in both having full outer shells.

In this example, the group 1 atoms lose an electron and the group 7 atoms gain an electron so each becomes electrically unstable, i.e. they form **ions**. Ions are charged particles formed when an atom gains or loses electrons.

Ionic bonding is common when **metals react with non-metals** (e.g. group 1 with group 7 and group 2 with group 6 elements). For example, sodium (metal) and chlorine (non-metal) react to form sodium chloride, an **ionic compound**. The ions formed are Na^+ and Cl^-.

H When forming an ionic compound, a **transfer of electrons** occurs. The metal atoms **lose electrons** and give them to the non-metal atoms. Each atom loses or gains enough electrons to form **full outer shells** for both ions.

Examples of ionic bonding can be seen in combinations involving:
● group 1 (alkali) metals with group 7 (halogens) non-metals e.g. NaCl
● group 2 (alkaline earth) metals with group 6 non-metals e.g. MgO.

> **Ion**: A charged particle produced as a result of an atom gaining or losing one or more electrons.
>
> **Ionic compound**: A compound formed between a metal and a non-metal in which the metal atom transfers electron(s) to the non-metal atom.

Sodium chloride (NaCl)

When sodium (electronic configuration: 2, 8, 1) reacts with chlorine (electronic configuration: 2, 8, 7), the single electron in sodium's third shell is **transferred** to the outer (third) shell of the chlorine (Figure 11.6).

The sodium has lost an electron so is no longer electrically neutral; it now has 11 protons and only 10 electrons (it lost an electron) so is represented by Na^+. The chlorine, having gained one electron, now has 18 electrons (but still 17 protons in the nucleus) so is electrically negative and is represented by Cl^-.

Because the Na^+ and the Cl^- are **oppositely charged ions**, they are **attracted to each other** to form the ionic compound sodium chloride. Note that the ions are not called 'chlorine' ions but chloride ions.

In the resulting compound, sodium chloride, both the sodium ions and the chloride ions now have full outer shells (sodium 2, 8 and chloride 2, 8, 8).

The strong attraction between oppositely charged ions in ionic bonding ensures that:
● ionic bonds are **strong**
● **substantial energy** is needed to break them.

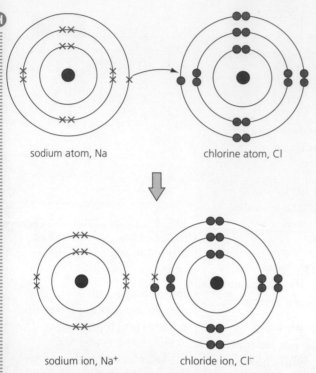

sodium atom, Na   chlorine atom, Cl

sodium ion, Na⁺   chloride ion, Cl⁻

Figure 11.6 Ion formation in sodium chloride

Covalent bonding

When **non-metal atoms** bond to form molecules, they **share** electrons (rather than transferring them from one atom to the other). This is to give every atom in the new molecule a full outer shell.

The shared electrons count in the electronic configuration of both atoms. The shared electrons make a **covalent bond** that holds the atoms together. Water (H_2O) and hydrogen chloride (HCl) are examples of **covalent compounds**.

A **hydrogen** molecule can be represented using a '**stick diagram**' in which each line in the diagram represents a shared pair of electrons (a single covalent bond): H–H.

Covalent bonding in hydrogen is illustrated in Figure 11.7.

Pairs of electrons not involved in bonding are described as **lone pairs** of electrons.

Figure 11.8 shows covalent bonding in water (H_2O), hydrogen chloride (HCl) and methane (CH_4).

> **Covalent bond**: A shared pair of electrons that holds two atoms together.
>
> **Covalent compound**: Two or more atoms joined together by sharing electrons (a covalent bond).

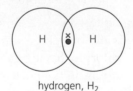

hydrogen, H_2

Figure 11.7 Covalent bonding in hydrogen

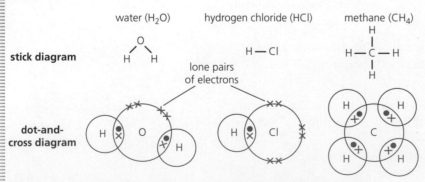

water (H_2O)   hydrogen chloride (HCl)   methane (CH_4)

stick diagram

lone pairs of electrons

dot-and-cross diagram

Figure 11.8 Covalent bonding in water, hydrogen chloride and methane

> **Exam tip**
>
> Using either stick (line) diagrams or dot-and-cross diagrams you can see that there are two covalent bonds in one molecule of water, one in hydrogen chloride and four in methane.

H Note that a molecule of water has two pairs of lone electrons and hydrogen chloride three pairs; there are no lone electrons in methane.

Now test yourself

6 Define the term 'ion'.
7 Name the type of bond produced when a shared pair of electrons holds two atoms together.

Answers on p. 151

Exam practice questions

1 Figure 11.9 shows the structure of an atom.
  (a) Name the parts labelled **A**, **B** and **C**. [3]
  (b) Give the atomic number of this atom. [1]
  (c) Give the mass number of this atom. [1]
  (d) Give the electronic structure of this atom. [1] **[6 marks]**

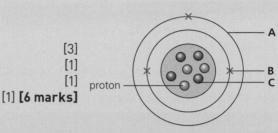

proton

Figure 11.9

2 Figure 11.10 shows an outline of the Periodic Table.

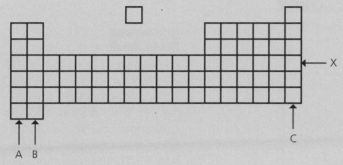

Figure 11.10

  (a) What do the small square boxes represent? [1]
  (b) Name the groups labelled **A**, **B** and **C**. [3]
  (c) Identify the number of period **X**. [1]
  (d) How many electrons are in the outer shell of an atom of a group **B** element? [1]

H 3 Write the balanced symbol equation for the reaction between potassium (an alkali metal) and water. [3]

4 Explain, in terms of ion formation, how sodium chloride is formed from sodium and chlorine. [5]

5 Figure 11.11 represents bond formation in a molecule of water.

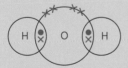

Figure 11.11

  (a) How many covalent bonds are shown in this diagram? [1]
  (b) How many lone pairs of electrons are shown? [1]
  (c) What is the evidence that covalent bonding is present? [2]

Answers online

12 Metals and the reactivity series

Not all metals are equally reactive. Metals can be placed in order of their reactivity in a series called the **reactivity series**.

The reactivity series

REVISED

The **most reactive metals** are those in **group 1** (the alkali metals).

Metals can be placed in the reactivity series based on how reactive they are with water and acid.

Table 12.1 shows the relative reactivity of a number of metals.

> **Reactivity series**: The order of metals according to how reactive they are with substances such as water or acid.

Table 12.1 The reactivity series

Metal	Symbol	Reactivity
Potassium	K	Most reactive
Sodium	Na	
Calcium	Ca	
Magnesium	Mg	
Aluminium	Al	
Zinc	Zn	
Iron	Fe	
Copper	Cu	Least reactive

> **Exam tip**
>
> The group 1 metals potassium and sodium are more reactive than the group 2 metals calcium and magnesium. Potassium is more reactive than sodium as it is further down group 1.

Metals reacting with water

The reactivity series can be based on:

- **metals reacting with cold water** – only the most reactive metals (potassium, sodium and calcium) will react with cold water

  Reaction

  metal + (cold) water → metal hydroxide + hydrogen

  Example

  sodium + water → sodium hydroxide + hydrogen

- **metals reacting with hot water or steam** – less reactive metals will react with hot water or, if very unreactive, steam (but not with cold water)

  Reaction

  metal + steam → metal oxide + hydrogen

  Example

  zinc + steam→ zinc oxide + hydrogen

Metals reacting with acid

The reaction of metals with dilute acid provides further evidence of where metals sit in the reactivity series.

Aluminium, zinc and iron will not react with cold water but will react with dilute acid:

Reaction

metal + dilute acid → salt + hydrogen

Example 1

zinc + sulfuric acid → zinc sulfate + hydrogen

Example 2

zinc + hydrochloric acid → zinc chloride + hydrogen

Copper, the least reactive metal in our series, does *not* react with water or acid.

By observation you should be able to place metals in order of reactivity. This can be done by comparing the rate at which bubbles of hydrogen are given off or the rate at which the metal gets used up in the reaction.

Prescribed practical C2

Investigate the reactivity of metals

Now test yourself

TESTED

1 Place the following metals in order of decreasing reactivity:
   calcium   aluminium   copper   sodium
2 Write the word equation for the reaction of aluminium and sulfuric acid.

Answers on p. 151

Energetics

REVISED

Exothermic reactions give out heat and **increase the temperature** of the surroundings. Most chemical reactions are exothermic. Examples of exothermic reactions include neutralisation reactions and the combustion of fuels.

Endothermic reactions take in heat from the surroundings and lead to a **decrease in temperature**.

Sports injury packs can be used to reduce swelling as a result of a sporting injury. These packs commonly contain two chemicals which are initially separated but which, when mixed, react endothermically and become very cold.

Exothermic: A reaction that gives out heat energy.

Endothermic: A reaction in which heat energy is taken in.

Prescribed practical C3

Investigate the temperature changes which occur during a reaction

Exam tip

In an exothermic reaction, the reactants have more energy than the products. In an endothermic reaction, the products have more energy than the reactants.

⊕Electrolysis

REVISED ☐

For **electrolysis** to take place, **carbon** or **graphite electrodes** are immersed in a liquid and connected to a power supply (Figure 12.1). A current flows between the negative electrode (**cathode**) and the positive electrode (**anode**). The liquid undergoing electrolysis must be able to conduct electricity; it must be an **electrolyte**. The electricity decomposes (breaks down) the electrolyte.

● Electrolytes have free ions that move and carry charge.
● The positive ions (**cations**) move to the cathode; the negative ions (**anions**) move to the anode.

The cations gain electrons to become atoms and the anions lose electrons to become atoms (they become electrically neutral).

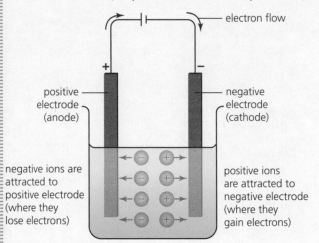

negative ions are attracted to positive electrode (where they lose electrons)

positive ions are attracted to negative electrode (where they gain electrons)

Figure 12.1 Electrolysis

Electrolysis: The process of using electricity to decompose an ionic compound.

Cathode: The negative electrode in electrolysis.

Anode: The positive electrode in electrolysis.

Electrolyte: The liquid that conducts electricity.

Cation: A positive ion.

Anion: A negative ion.

Exam tip

Carbon and **graphite** are suitable to be used as electrodes as they are **inert** (unreactive).

Extracting aluminium from alumina (aluminium oxide)

The method of extraction of aluminium using electrolysis (Figure 12.2) is summarised below.

1 The aluminium ore (bauxite) is purified to form aluminium oxide (alumina).
2 The aluminium oxide is dissolved in molten cryolite, which lowers the temperature needed to carry out the process (saving energy bills).
3 An aluminium oxide crust keeps the heat in (the process operates at 900–950 °C).
4 At the negative electrode, positive aluminium ions combine with electrons to form aluminium metal.
5 At the positive electrode, negative oxide ions form oxygen gas.

H The ionic (half) equation for the reaction at the **cathode** is:

$$Al^{3+} + 3e^- \rightarrow Al$$

Explanation of reaction: the aluminium cations (Al^{3+}) gain three electrons ($3e^-$) to become aluminium atoms.

The ionic (half) equation for the reaction at the **anode** is:

$$2O^{2-} - 4e^- \rightarrow O_2$$

Explanation of reaction: the oxide anions lose two electrons at the anode to become oxygen atoms.

The **carbon anode has to be replaced** periodically as it wears away because of it reacting with oxygen:

$$C + O_2 \rightarrow CO_2$$

Explanation of reaction: the C comes from the electrode and the O_2 is produced as a consequence of the oxygen formed during the electrolysis.

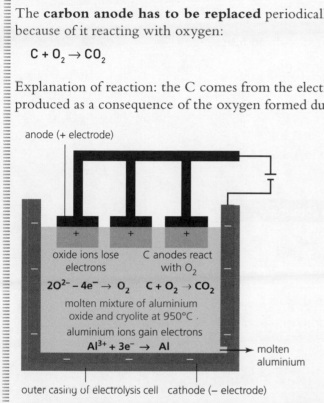

Figure 12.2 The extraction of aluminium by electrolysis

Although the use of cryolite saves heat energy, the process is still very expensive and the extraction of aluminium oxide from bauxite leaves a lot of waste. Consequently, it is much more cost efficient to **recycle aluminium** where possible.

Now test yourself

TESTED ☐

3 In electrolysis, describe what happens to the positive ions.
4 In the manufacture of aluminium, give the ionic (half) equation for what happens to the aluminium ions in this reaction.

Answers on p. 151

Flame tests

Metal ions produce a coloured flame when heated in a Bunsen flame. Different metals produce different coloured flames. The colour of the flame can be used to identify the metal in a sample; this is called a **flame test**. Flame tests are used to identify metal ions found at the scene of a crime. They can be compared to those found on a suspect's clothing. **Chloride** compounds containing the metal ions will give the colours described in Table 12.2.

> **Flame test**: A test used to identify some metal ions by the flame colour they produce when burning.

Safety: It is important to wear eye protection and to take care when using a Bunsen burner.

These are the steps for carrying out a flame test:
1 Clean the flame-test rod (nichrome wire) in concentrated acid and heat it in a blue Bunsen flame.
2 Dip the end of the cleaned rod in the sample solution (containing the metal ions) and then hold the rod in the flame.
3 Observe and record the colour of the flame produced.
4 The rod must be cleaned between samples.

The flame test colours of a range of metal ions are shown in Table 12.2.

Table 12.2 Flame tests

Metal	Metal ion	Flame colour
Lithium	Li^+	Crimson
Sodium	Na^+	Yellow-orange
Potassium	K^+	Lilac
Calcium	Ca^{2+}	Brick-red
Copper	Cu^{2+}	Blue-green

Exam practice questions

1 (a) Place the following four elements in order of decreasing reactivity:
   sodium   copper   zinc   potassium   [1]
  (b) Write the word equation for potassium reacting with water.   [2]
  (c) Write a balanced symbol equation for magnesium reacting with hydrochloric acid. [3]
2 (a) (i) What is meant by an 'exothermic' reaction?   [1]
      (ii) Give an example of an exothermic reaction.   [1]
  (b) What is the evidence that Figure 12.3 represents an endothermic reaction?   [1]

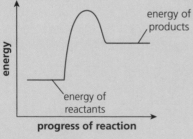

Figure 12.3

Exam practice answers at **www.hoddereducation.co.uk/myrevisionnotesdownloads**

H

3 **(a)** What is meant by the term 'electrolyte'? [1]
 (b) Explain fully the function of cryolite in the extraction of aluminium. [2]
 (c) Explain fully why the carbon anodes have to be replaced periodically. [3]
 (d) Write the ionic equation for the reaction that occurs at the cathode. [3]
4 Copy and complete Table 12.3 about flame tests. [3]

Table 12.3

Metal	Flame colour
Lithium	_____ [1]
_____ [1]	Yellow-orange
Potassium	_____ [1]

Answers online

ONLINE

13 Materials

Natural and synthetic materials and their properties

There are many different materials. They can be broadly grouped as:
- **natural materials** – these are not made by processes involving chemical methods and include examples such as granite and wood; they can be obtained from living things such as cotton, wool and silk
- **synthetic (man-made) materials** – these involve being processed by chemical methods and include plastics and glass.

Different materials differ from each other in many ways, including:
- melting point
- boiling point
- strength
- conductivity
- density
- hardness.

> **Natural material**: A material that can be obtained from living things, such as wool, or made without being processed by chemical methods, such as granite.
>
> **Synthetic material**: A man-made or manufactured material, for example plastic.

Exam tip

You should be aware that materials can differ from each other in a number of ways and not just the **physical properties** in the bullet points above. They can differ in cost and appearance, for example.

Some of the main types of material and their properties are summarised in Table 13.1.

Table 13.1 Properties of materials

	Metals	Plastics	Fibres	Ceramics
Properties	<ul><li>Strong, hard and high density; not flexible</li><li>Good conductors of heat and electricity</li><li>High melting point</li></ul>	<ul><li>Strong and hard or flexible (depending on plastic)</li><li>Low density</li><li>Poor conductors of electricity or heat</li><li>Low melting point</li></ul>	<ul><li>Flexible (not strong or hard)</li><li>Low density</li><li>Poor conductors of electricity</li><li>Low melting point</li></ul>	<ul><li>Strong, hard</li><li>High density</li><li>Not flexible</li><li>Unreactive</li><li>Poor conductors of electricity</li><li>High melting point</li></ul>
Uses	Building; kitchen utensils; electrical wiring; engines	Making chairs; bottles; window frames; doors; clingfilm; covers for electronic devices	Clothes; curtains; carpets	Bricks for building; crockery; cups
Examples	Copper; iron; aluminium	PVC; polythene; Bakelite	Cotton; silk; nylon	Pottery; china

Modern synthetic (man-made) materials have many advantages over traditional materials. They usually:
● are **cheaper** (and easier to process)
● have **'better' properties**.

For example, nylon has been used to replace linen in the manufacture of tablecloths and napkins. Linen takes a long time to make and involves many complex processes to convert the flax plants into finished products. In addition, linen creases very easily and is difficult to iron.

This has led to the decline of the Northern Ireland linen industry; an industry that once employed many thousands of people.

Smart materials

Smart materials change their properties if exposed to changes in **the environment (their surroundings)**.

Examples of smart materials include:
● **thermochromic** paints and dyes, which change colour when the **temperature** changes, e.g. colour-changing T-shirts
● **photochromic** paints and dyes, which change colour when exposed to **light** or changes in light intensity, e.g. sunglasses that become darker in stronger sunlight.

Nanomaterials contain very small **nanoparticles**.

Uses of nanomaterials include:
● **sun creams** – the nanoparticles give better **protection from UV rays** and better **skin coverage** when added; however, they have some possible risks including **skin damage** and they can be **harmful to the environment**
● **wound dressings** and in **sterilising sprays**.

> **Smart material**: A material that changes its properties if exposed to a particular environmental change.
>
> **Thermochromic**: The term given to paints or dyes that change their colour when their temperature changes.
>
> **Photochromic**: The term given to paints and dyes (or other materials) that change their colour as light intensity changes.
>
> **Nanoparticles**: Structures that contain a few hundred atoms; typically 1–100nm in size.
>
> **Nanomaterial**: A material made from nanoparticles.

Now test yourself

TESTED

1 Give **one** advantage of using synthetic materials compared with natural materials.
2 Give **one** example of a synthetic material.
3 Name the type of 'smart' material that changes colour when in different temperatures.

Answers on p. 151

Ⓗ Nanomaterials and emergent materials

Nanotechnology

Nanotechnology involves technology using extremely small particles (**nanoparticles**); the nanoscale is around 10^{-9} metres, so 1nm = 0.000000001m. A nanomaterial particle contains **a few hundred atoms** and is in the range **1–100nm**.

Nano-sized materials have **different properties** compared with 'normal-sized' materials.

Graphene

Graphene is a new (emerging) material. It consists of a single layer of **graphite**; in other words, a layer **one atom thick**.

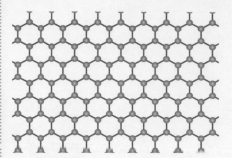

Figure 13.1 Graphene is a single layer of graphite

Graphene is **very strong** and can **conduct electricity**. It has many uses such as in **batteries**, **solar cells** and in **electronics** (e.g. touchscreens). Recently it has been shown that it can be used as a molecular sieve to filter the salt out of sea water.

Fullerenes and other emerging carbon-based materials

The molecule C$_{60}$ (Figure 13.2) is an example of a **fullerene**. Fullerenes are carbon atoms arranged as **hexagonal rings** (rings of six atoms), or rings of five or seven atoms. However, they all have a hollow centre.

Figure 13.2 C$_{60}$ has a shape that resembles a football

> **Fullerens**: A family of carbon molecules each with carbon atoms linked in rings to form a hollow sphere or tube.

> **Exam tip**
>
> The molecule C$_{60}$ is often referred to as a '**buckyball**'.

Fullerenes are used to **deliver medicines** into the body and they are also being used in engineering as **lubricants**.

Carbon nanotubes (Figure 13.3) are **cylindrical fullerenes** (sometimes called 'buckytubes'); they are effectively long tubes of graphene. They have **very high strength** and can **conduct electricity**. Carbon nanotubes are used to make high-quality sports equipment such as **golf clubs**.

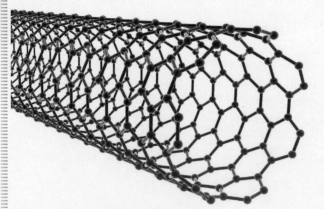

Figure 13.3 A carbon nanotube

Exam practice answers at **www.hoddereducation.co.uk/myrevisionnotesdownloads**

Using only the information provided, suggest why fullerenes are suitable for use in delivering medicines into the body and for use as lubricants (reducing friction between moving parts in machines).

Answer

- Medicines can be carried in the hollow core of the fullerene.
- The spherical shape of the fullerene allows them to roll over and past each other, reducing friction between moving parts.

Using materials to fight crime

REVISED

Forensic evidence

Forensic evidence is evidence that can be used in investigating crime. Forensic evidence is often collected at a crime scene. The main types of forensic evidence are summarised in Table 13.2.

Table 13.2 Types of evidence

Type of evidence	Examples
Biological	Includes blood, sweat, skin tissue, semen; biological evidence can be used to give a genetic fingerprint; everyone has a unique genetic fingerprint
Fingerprint	Can be taken from a crime scene and compared with a suspect's fingerprint
Footprints or tyre tracks	Often not conclusive; many cars can have the same tyres
Trace evidence	Hair, fibres from materials, paint or glass fragments; fibres from clothes or paint from a car can be left at a crime scene and later identified and compared with a suspect's, e.g. using flame tests
Digital	Evidence on phones and computers, including internet logs and emails
Drug	Blood analysis can show if drugs are being used
Explosive	Chemicals on the skin can show whether explosive materials were handled

Trace evidence: A small amount of evidence left at a crime scene, such as hair or fibres from clothing. Trace evidence normally needs scientific analysis.

Fingerprint evidence

Everyone has a pattern of fine lines (a **fingerprint**) on their fingertips. There are four main types of fingerprint pattern: **arch**, **loop**, **whorl** and **composite** (Figure 13.4).

Fingerprint: The unique pattern of lines found on a person's fingertips.

Figure 13.4 The four types of fingerprint

Every person has a **unique** fingerprint pattern (no two people have the same fingerprint). This is what makes fingerprinting so valuable in criminal trials.

If you press your fingertips against a hard surface, you leave a fingerprint on that surface. Scientific techniques are needed to get a fingerprint and make it visible. There are several stages involved:

1 The fingerprint marks have to be **dusted** with a powder. If the surface is **white**, **carbon black** powder is used. If the surface is **black** or **mirrored**, **aluminium dust** is used.

2 Once dusted, the fingerprint is **transferred**. The excess dust is brushed off and sticky tape is pressed over the fingerprint. The dust sticks to the sticky tape and it can then be transferred to a card. Carbon prints are transferred to white card (for contrast) and aluminium powder prints are transferred to black card.

3 The fingerprints can be **compared** with the previously taken fingerprints of suspects. For over 100 years fingerprints have been used in the court system. Because fingerprints are **unique**, they can be used to help prove both guilt and innocence.

Alternative light sources and **chemical developers** can be used to help obtain fingerprints. For example, **UV light** can be used or normal (white) light with **special filters** added. Chemical developers, including fluorescent dyes, can be used to help make the fingerprints show up more. Alternative light sources and chemical developers can make the fingerprint clear enough to be **photographed** or **scanned**.

Forensic scientists can also use a photograph or scan of a fingerprint to compare with a **database** of fingerprints to help identify individuals. This is also a scientific process in that many features of the fingerprint obtained are compared with the same features stored in the database.

Fingerprint recognition systems

Fingerprint recognition systems are becoming increasingly common. The principle involves individuals having their fingerprints scanned and stored. They are important **security systems** for the protection of **mobile phones**, **tablets** and other expensive equipment. In many schools, attendance is monitored using fingerprint recognition systems.

> **Exam tip**
>
> Alternative light sources and chemical developers can be used to obtain fingerprints from surfaces that are not suitable for traditional techniques, such as clothing and skin. They are also usually a much quicker method.

> **Exam tip**
>
> Fingerprint recognition systems work because everyone has a unique fingerprint.

Now test yourself

4 What is meant by the term 'digital evidence'?
5 Name the four types of fingerprint.

Answers on p. 151

Exam practice questions

H

1 (a) What is meant by the term 'synthetic material'? [1]
   (b) Give **two** examples of synthetic materials. [2]
2 (a) Describe what is meant by a nanomaterial. [2]
   (b) (i) Give **two** properties of graphene. [2]
       (ii) Give **one** function of graphene. [1]
3 Describe how you would take a fingerprint from the door of a black car. [4]
4 (a) Fingerprint recognition systems are becoming increasingly common as systems for monitoring attendance in schools.
       (i) Suggest **two** advantages of using fingerprint recognition systems in schools rather than the traditional register method. [2]
       (ii) Suggest **one** disadvantage. [1]
   (b) Give **one** other way in which fingerprint recognition systems are used. [1]

Answers online

ONLINE

14 Rates of reaction

The **rate of reaction** is the speed at which a reaction is taking place.

Measuring the rate of reaction

The rate of a reaction may be determined by measuring the rate at which a **reactant is used** or a **product is formed over time**.

$$\text{mean rate of reaction} = \frac{\text{quantity of reactant used}}{\text{time}} \text{ or } \frac{\text{quantity of product produced}}{\text{time}}$$

> **Rate of reaction**: The amount of reactant used or product formed per unit time.

Measuring reactant used and product produced

A **gas syringe** can be used to record the volume of a gas produced over time (Figure 14.1).

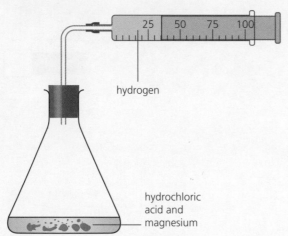

Figure 14.1 A gas syringe

> **Exam tip**
>
> **Mass** (g) of a chemical or **volume** (cm^3) of a gas are normally used to measure the change in a reactant or product over time.

A **balance** can be used to record change in a product formed (Figure 14.2), but can also be used to record change in gas produced over time (measured as a decrease in mass over time).

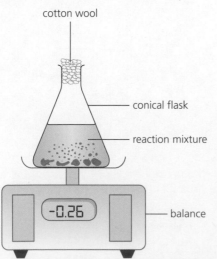

Figure 14.2 Using a balance to record mass of gas produced

> **Exam tip**
>
> In Figure 14.2 the cotton wool in the neck of the flask allows the gas produced to escape and prevents any reactants from splashing out.

Reaction rate graphs

By measuring the quantity of reactant used or product produced, it is possible to draw graphs showing the rate of a reaction. The slope of the graph represents the rate of reaction. Figure 14.3 shows a typical rates of reaction graph.

Figure 14.3 A typical rates of reaction graph

Now test yourself

TESTED

1 Copy and complete the sentence.
   When measuring rate of reaction, you can either measure the rate of _____ used or the rate of _____ formed over time.
2 In a rates of reaction graph, what does it mean when the graph line gets steeper?

Answers on p. 151

Measuring rates of reactions in metals and metal carbonates when reacting with dilute acid

Metals react with **dilute acid** to produce the salt of the metal and **hydrogen**. The rate of reaction between a metal, such as magnesium, and sulfuric acid can be measured by recording the volume of hydrogen produced using a gas syringe or a balance, as described previously.

Equation:

> magnesium + sulfuric acid → magnesium sulfate + hydrogen

Worked example

In the reaction of magnesium with sulfuric acid, 0.20 g of magnesium was used up in 50 seconds. What is the mean rate of reaction?

Answer

$$\text{rate of reaction} = \frac{\text{mass of reactant used}}{\text{time}} = \frac{0.20}{50} = 0.004\,\text{g/s}$$

Metal carbonates react with **dilute acid** to produce the **salt of the metal**, **water** and **carbon dioxide**. With these reactions, the rates of the reaction can be measured by recording the volume of carbon dioxide produced over time.

Equation:

calcium carbonate + hydrochloric acid → calcium chloride + water + carbon dioxide

Worked example

In the reaction of calcium carbonate with hydrochloric acid, $20\,cm^3$ of carbon dioxide was produced in 40 seconds. What is the mean rate of reaction?

Answer

$$\text{rate of reaction} = \frac{\text{volume of gas produced}}{\text{time}} = \frac{20}{40} = 0.5\,cm^3/s$$

Factors affecting the rate of reaction

The rate of reaction is affected by **temperature** and the **concentration of the reactants**.
- The higher the temperature, the faster the reaction.
- The higher the concentration of reactants, the faster the reaction.

H Higher temperatures produce faster reactions as the particles have **more energy** and they will be **moving faster** so will **collide more often**. Higher temperatures also give the molecules more energy to react.

The higher the concentration of the reactants, the faster the rate of reaction as the reactant particles are **closer together** and are therefore **more likely to collide** and react.

Catalysts

A **catalyst** is a substance that **increases the rate of reaction** without being used up.

For example, hydrogen peroxide decomposes to form water and oxygen. The addition of manganese(IV) oxide acts as a catalyst in speeding this reaction up.

Catalyst: A substance which causes a reaction to go faster without being used up during the reaction.

Exam tip

The **catalyst is not used up** in the reaction; there will be as much catalyst present at the end of the reaction as there was at the start.

Exam practice questions

1 (a) Describe what is meant by the rate of a reaction. [1]
  (b) Name **two** pieces of apparatus that can be used to measure the rate at which a gas is produced in a reaction. [2]
2 Calcium carbonate reacts with hydrochloric acid to produce calcium chloride, water and carbon dioxide. In a reaction, $140\,cm^3$ of carbon dioxide was produced in 1 minute 10 seconds. Calculate the mean rate of reaction during this time in cm^3/s. [2]

Exam practice answers at **www.hoddereducation.co.uk/myrevisionnotesdownloads**

3 **(a)** Figure 14.4 is a typical rates of reaction graph showing the mass of gas formed in three reactions. The same mass of reactant was used in each reaction. Give **two** pieces of evidence that show that reaction **X** is the fastest reaction. [2]

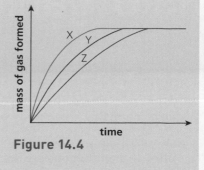

Figure 14.4

(b) Figure 14.5 shows how the mass of a reactant decreases during a reaction.

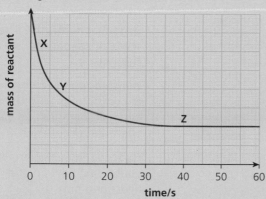

Figure 14.5

    (i) At which point (**X**, **Y** or **Z**) was the reaction occurring fastest? [1]

    (ii) At what time was the reaction finished? [1]

4 **(a) (i)** Explain the role of a catalyst in a chemical reaction. [2]

    (ii) Give **one** example of a catalyst. [1]

  (b) Describe and explain the effect of temperature on reaction rates. [3]

Answers online

ONLINE ☐

15 Organic chemistry

Organic chemistry is the study of compounds containing **carbon**.

There are so many carbon compounds that they are grouped into 'families' or **homologous series**.

Two homologous series are the **alkanes** and the **alkenes**.

Hydrocarbons

REVISED

Hydrocarbons are a very important group of carbon-containing compounds. Hydrocarbons contain the elements **carbon** and **hydrogen** *only*.

The alkanes

The **alkanes** are a homologous series of **saturated hydrocarbons** with the general formula C_nH_{2n+2}.

The molecular and structural formulae of the first four alkanes are shown in Figure 15.1.

Name	Molecular formula	Structural formula
Methane	CH_4	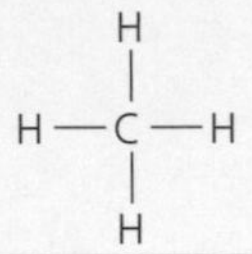 $H-\underset{\displaystyle H}{\overset{\displaystyle H}{C}}-H$
Ethane	C_2H_6	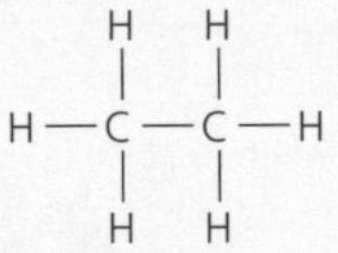 $H-\underset{H}{\overset{H}{C}}-\underset{H}{\overset{H}{C}}-H$
Propane	C_3H_8	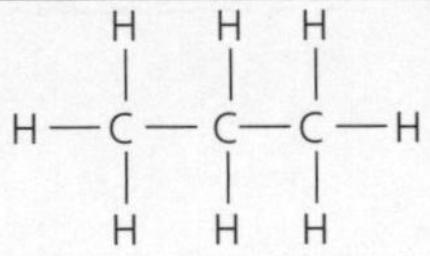 $H-\underset{H}{\overset{H}{C}}-\underset{H}{\overset{H}{C}}-\underset{H}{\overset{H}{C}}-H$
Butane	C_4H_{10}	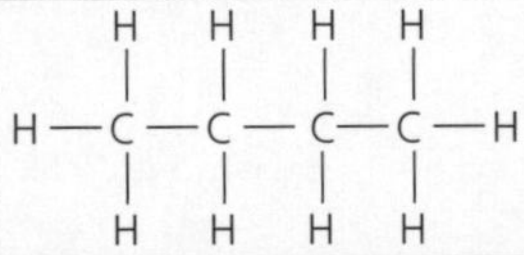 $H-\underset{H}{\overset{H}{C}}-\underset{H}{\overset{H}{C}}-\underset{H}{\overset{H}{C}}-\underset{H}{\overset{H}{C}}-H$

Figure 15.1 The first four alkanes

Each of the alkanes in the table is a **gas** at room temperature and pressure.

> **Organic chemistry:** The study of carbon-containing compounds.
>
> **Homologous series:** A group of compounds with the same general formula and similar chemical properties.
>
> **Hydrocarbon:** A compound containing carbon and hydrogen only.
>
> **Alkanes:** A group (homologous series) of saturated hydrocarbon compounds with the general formula C_nH_{2n+2}

> **Exam tip**
>
> 'Saturated' means that there are no double covalent bonds.

> **Worked example**
>
> The general formula for the alkanes is C_nH_{2n+2}. Using this information, work out the molecular formula of propane, an alkane with three carbon atoms.
>
> *Answer*
>
> Hydrogen is the only other element present and, in the formula, $n = 3$.
>
> number of hydrogen atoms = $(2 \times 3) + 2 = 8$
>
> So the formula is C_3H_8.

Crude oil

Crude oil is formed from plants and animals that died **millions of years** ago. **Pressure** and **heat** changed their remains into crude oil. It is a liquid mixture containing hundreds of different substances. As well as liquids, crude oil contains some gases and dissolved solids. Crude oil is the main source of the hydrocarbons that are used in the world today.

> **Crude oil**: A fossil fuel; a liquid mixture containing many different substances that can be separated by fractional distillation.

Fractional distillation

Because crude oil is a mixture of substances, the different compounds can be separated by **fractional distillation** (Figure 15.2). This process involves the following stages:

● The crude oil is heated.
● On heating, the oil **evaporates**, causing most of it to become a mixture of gases.
● The gases move up the fractioning column (steel tower) used for fractional distillation.
● As each fraction reaches its boiling point, it is separated off from the mixture and **condenses** back into a liquid.
● Because the tower is cooler at the top than the bottom, **different liquids condense at different levels** in the tower; the liquids are collected separately.
● The **solids** dissolved in crude oil (bitumen/tar) drain out of the bottom of the tower (they do not evaporate).The **gases** do not condense; they come out of the top of the tower as refinery gas.

There are three key points about the fractional distillation of crude oil:
● The different components produced are called **fractions**.
● All the fractions are **hydrocarbons** and most are **alkanes**.
● Each fraction (separated at a particular position in the tower) has a **similar number of carbon atoms**. The smaller hydrocarbons with smaller numbers of carbon atoms in their molecules form the fractions at the top of the tower.

> **Exam tip**
>
> Crude oil is a **finite resource**. This means that there is only a certain amount left and the reserves will run out over time.

> **Fractional distillation**: The process used to separate the compounds in crude oil. It involves heating and evaporation, and the subsequent condensation of gases back to liquids.

> **Exam tip**
>
> The term '**fractional distillation**' summarises what happens. 'Distillation' involves heating liquids, causing them to evaporate into gas and then condensing them back into liquids as they reach a cooler zone. 'Fractional' means that this happens to the different substances in different places because they have **different boiling points**.

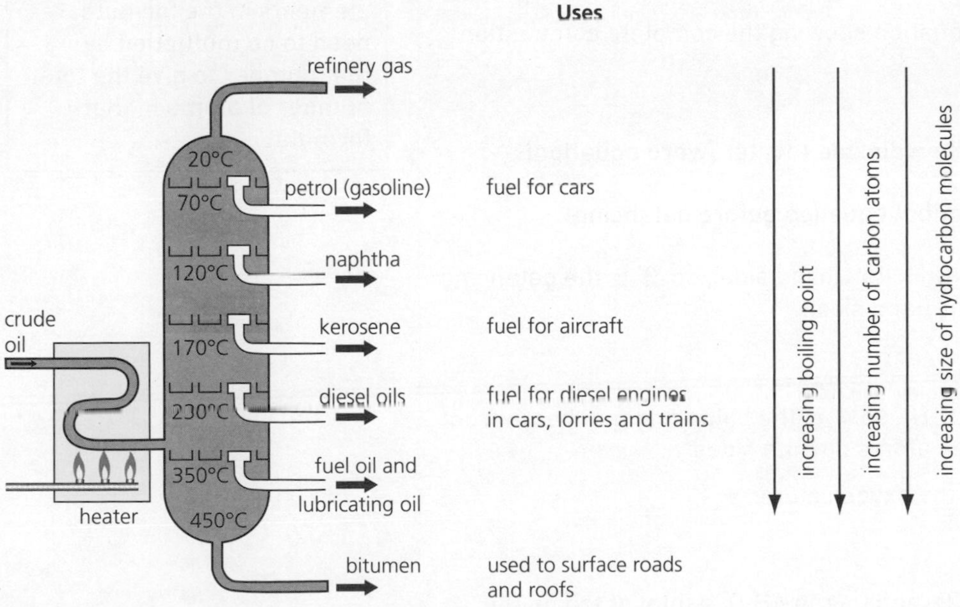

Figure 15.2 The fractional distillation of crude oil

Now test yourself

TESTED

1 Give the general formula for an alkane.
2 State **one** way in which bitumen is used.

Answers on p. 151

Exam tip

You need to know the names and uses of the following fractions: petrol, kerosene, diesel and bitumen.

Combustion of alkanes

All hydrocarbons can be burned; some more easily than others. When hydrocarbon fuels are combusted in air, the hydrogen reacts with the oxygen to form water and the carbon reacts with the oxygen to form carbon dioxide. Combustion also releases heat energy, although this is not shown in equations:

hydrocarbon + oxygen → carbon dioxide + water

For example:

methane + oxygen → carbon dioxide + water

propane + oxygen → carbon dioxide + water

Ⓗ The balancing of the symbol equation for the combustion of methane is as follows:

$CH_4 + O_2 \rightarrow CO_2 + H_2O$ (symbol equation before balancing)

$CH_4 + 2O_2 \rightarrow CO_2 + 2H_2O$ (balanced symbol equation)

The equation had to be balanced by writing '2' in front of 'O$_2$' and '2' in front of 'H$_2$O' to get four O atoms and four H atoms on both sides of the equation.

Exam tip

The equations shown for the combustion of alkanes are those for their **complete combustion**. You are only required to know equations involving their complete combustion. Complete combustion takes place when there is a good supply of oxygen.

Complete combustion: The burning of a fuel when there is a good supply of oxygen.

> ### Worked example
>
> Write a balanced symbol equation showing the complete combustion of propane.
>
> *Answer*
>
> propane + oxygen → carbon dioxide + water (word equation)
>
> $C_3H_8 + O_2 \rightarrow CO_2 + H_2O$ (symbol equation before balancing)
>
> There are three C atoms in C_3H_8 (left-hand side), so '3' is the balancing number in front of CO_2 (right-hand side):
>
> $C_3H_8 + O_2 \rightarrow 3CO_2 + H_2O$
>
> There are eight H atoms in C_3H_8, so '4' is the balancing number in front of H_2O to give eight hydrogen atoms on both sides.
>
> The next step is to balance the oxygen atoms:
>
> $C_3H_8 + O_2 \rightarrow 3CO_2 + 4H_2O$
>
> There are six O atoms in $3CO_2$ and four in $4H_2O$; a total of ten on the right-hand side. Multiplying O_2 on the left-hand side by 5 gives the balanced symbol equation:
>
> $C_3H_8 + 5O_2 \rightarrow 3CO_2 + 4H_2O$

Exam tip

Remember that when a number is put in front of a formula (when balancing equations), then *all* the elements in the formula need to be multiplied by that number to give the total number of atoms in that formula.

⊕ The alkenes

The **alkenes** are a homologous series of **unsaturated hydrocarbons**. They are unsaturated because they contain a double covalent bond between two carbon atoms.

The general formula of an alkene is C_nH_{2n}. This means that in the alkenes there are twice as many hydrogen atoms as carbon atoms in each molecule.

Figure 15.3 gives the names, molecular formulae and structural formulae of ethene, propene and butene, which are the first three members of the alkenes.

Name	Molecular formula	Structural formula
Ethene	C_2H_4	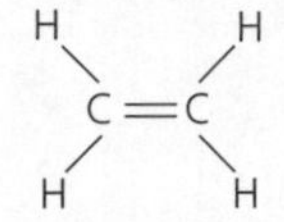
Propene	C_3H_6	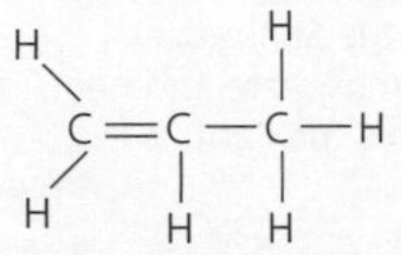
Butene	C_4H_8	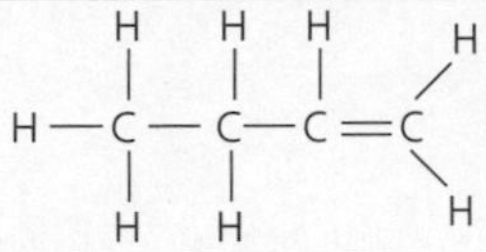

Figure 15.3 The first three alkenes

Ethene, propene and butane are **gases** at room temperature and pressure.

Combustion of alkenes

As with the alkanes, the **complete combustion** of alkenes produces carbon dioxide and water:

> alkene + oxygen → carbon dioxide + water

For example, when ethene burns it forms carbon dioxide and water, releasing heat:

> ethene + oxygen → carbon dioxide + water

The balanced symbol equation for the complete combustion of ethene is:

$$C_2H_4 + 3O_2 \rightarrow 2CO_2 + 2H_2O$$

The balanced symbol equations for the complete combustion of propene and butene are:

propene:

$$2C_3H_6 + 9O_2 \rightarrow 6CO_2 + 6H_2O$$

butene:

$$C_4H_8 + 6O_2 \rightarrow 4CO_2 + 4H_2O$$

Alkenes: A group (homologous series) of unsaturated hydrocarbon compounds with the general formula C_nH_{2n}.

Exam tip

There is no alkene with one carbon atom (i.e. no meth**ene**) as alkenes have a double bond between **two carbon** atoms.

Exam tip

Note that in Figure 15.3 there is one double bond in each of the structural formulae and that each carbon atom has four bonds.

Exam tip

The **alkanes** all end with **-ane** and the **alkenes** end with **-ene**.

Now test yourself

3 Alkenes are unsaturated hydrocarbons. What does the term 'unsaturated' mean?
4 Give the general formula for the alkenes.
5 Give the general equation for the complete combustion of an alkane or an alkene.

Answers on p. 151

TESTED

Atmospheric pollution

The combustion of hydrocarbons (as fossil fuels) is a major source of **atmospheric pollution**. As a consequence of a massive increase in the combustion of fossil fuels over the last two centuries, the concentration of carbon dioxide in the atmosphere has increased considerably.

Carbon dioxide is a **greenhouse gas**, trapping heat from the Sun within the Earth's atmosphere (the '**greenhouse effect**') and causing global temperatures to rise; a process known as **global warming** (Figure 15.4).

> **Greenhouse effect**: The heating effect caused by a layer of greenhouse gases trapping heat within the Earth's atmosphere.

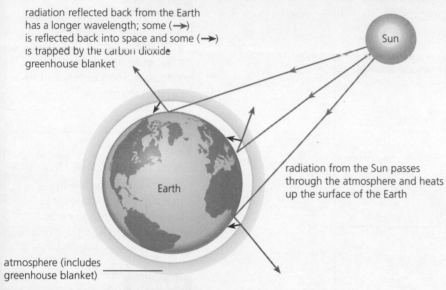

radiation reflected back from the Earth has a longer wavelength; some (→→) is reflected back into space and some (→→) is trapped by the carbon dioxide greenhouse blanket

Sun

radiation from the Sun passes through the atmosphere and heats up the surface of the Earth

Earth

atmosphere (includes greenhouse blanket)

Figure 15.4 The greenhouse effect

The causes and effects of global warming are summarised in Figure 15.5.

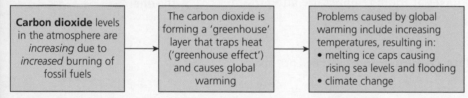

| **Carbon dioxide** levels in the atmosphere are *increasing* due to *increased* burning of fossil fuels | The carbon dioxide is forming a 'greenhouse' layer that traps heat ('greenhouse effect') and causes global warming | Problems caused by global warming include increasing temperatures, resulting in:
• melting ice caps causing rising sea levels and flooding
• climate change |

Figure 15.5 Global warming

⊕Polymers

Many small molecules (**monomers**) can be joined together to make **polymers** (Figure 15.6) in a process called **polymerisation**.

> **Monomer**: A small molecule that can be joined together with many other molecules to make a polymer.
>
> **Polymer**: A large molecule produced by the joining together of many monomers in a process called polymerisation.

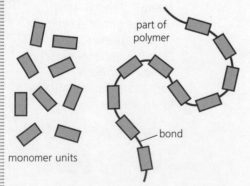

part of polymer

bond

monomer units

Figure 15.6 Monomers and their polymer

Exam practice answers at **www.hoddereducation.co.uk/myrevisionnotesdownloads**

H One type of polymerisation is **addition polymerisation** where the monomers have a carbon–carbon double bond and the monomers are basically 'added together'.

Two common polymers, built up by addition polymerisation, and their monomers are shown in Table 15.1.

Table 15.1 Polymers and polymerisation

Polymer	Monomer sub-unit
Polythene	Ethene
PVC (polyvinyl chloride)	Chloroethene (vinyl chloride)

The same basic principles apply when writing symbol (structural) equations for polymerisation reactions (Figure 15.7). There are a few key points:
- The monomer must have a double bond (e.g. C=C).
- We start with 'n' molecules of the monomer; n is a large number.
- The polymer has a single bond between the two carbon atoms.
- The polymer is written in square brackets and has 'n' after it to show that the polymer repeats what is shown in the brackets n times.

The equations for addition polymerisation of ethene and chloroethene (vinyl chloride) are shown in Figure 15.8.

if* = H then it makes ethene;
if* = Cl then it makes vinyl chloride (chloroethene)

point where additional monomer molecules join

number of monomer molecules used in the reaction

the monomer has a double (=) bond – one bond is broken to allow the monomers to join together to form the polymer

the polymer has single bonds between carbon atoms

the number of times the basic structure in brackets is repeated

Figure 15.7 The general reaction for polymerisation

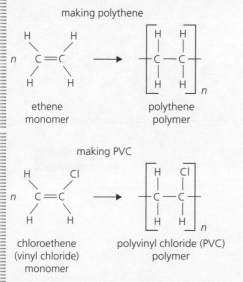

making polythene

ethene monomer

polythene polymer

making PVC

chloroethene (vinyl chloride) monomer

polyvinyl chloride (PVC) polymer

Figure 15.8 Polymerisation of ethene and chloroethene (vinyl chloride)

Addition polymerisation: A reaction in which small molecules (monomers) are joined or added together to make a long-chain molecule (a polymer).

Exam tip

Polymerisation equations are commonly asked for in exams. You are often given the monomer structure and asked to complete polymerisation. Do not forget to add the 'n' and change the double bond to a single bond between the two C atoms. It is also important to show the horizontal lines extending through the brackets.

Now test yourself

6 Name the monomer that forms the polymer polythene.
7 State **one** way in which the bonding in this monomer is different from the bonding in polythene.

Answers on p. 151

TESTED

Problems with waste plastics

Addition polymers (plastics) are **non-biodegradable**. Non-biodegradable means that they **cannot be broken down** by microorganisms (bacteria and fungi). Many of these non-biodegradable plastics are disposed of by **incinerating** (burning) or are dumped in **landfill**.

The advantages and disadvantages of incineration and landfill are summarised in Table 15.2.

Biodegradable: A substance that can be decomposed (broken down) by microorganisms.

Non-biodegradable: A substance that cannot be decomposed (broken down) by microorganisms.

Table 15.2 The advantages and disadvantages of incineration and landfill

Method	Advantages	Disadvantages
Landfill	● Convenient in that bin lorries simply dump their rubbish ● Once the landfill site is full it can be covered over and used for other purposes	● There is a shortage of landfill sites ● They produce lots of methane (a greenhouse gas) ● They are unsightly ● They can leach harmful chemicals into waterways
Incineration	● Takes up very little space	● Burning plastics can produce toxic and harmful fumes ● Carbon dioxide is released into the air, which adds to the greenhouse effect

Both landfill and incineration have more disadvantages than advantages; where possible it is best to recycle plastic.

Exam practice questions

1 Explain the term 'hydrocarbon'. [2]
2 (a) Describe the process of fractional distillation. [3]
  (b) Name **three** products of the fractional distillation of crude oil. [3]
  (c) Give **one** use for **one** of the products. [1]
3 The molecular formula for ethane is C_2H_6.
  (a) Draw the structural formula of ethane. [1]
  (b) Copy and complete the balanced symbol equation for the complete combustion of ethane: [2]

$$C_2H_6 + O_2 \rightarrow$$

4 (a) Describe the process of polymerisation. [2]
  (b) Copy and complete the structural equation to show how ethene is made into polythene: [3]

$$n \quad \begin{array}{c} H \quad H \\ | \quad | \\ C=C \\ | \quad | \\ H \quad H \end{array} \longrightarrow$$

ethene monomer

Answers online

ONLINE

16 Electrical circuits

Electrical circuits and symbols

Electricity can flow in an electric circuit. **Circuit diagrams** can be used to illustrate electric circuits. Figure 16.1 shows the symbols used to represent the main components in a circuit diagram.

Component	Symbol	Function
Battery		To supply electricity
Cell	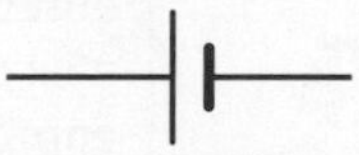	To supply electricity
Bulb	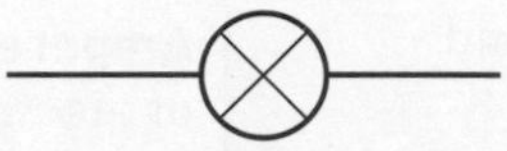	To convert current to light
Switch		To control the flow of current
Fuse	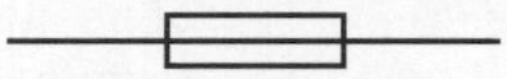	To stop too much current flowing
Voltmeter	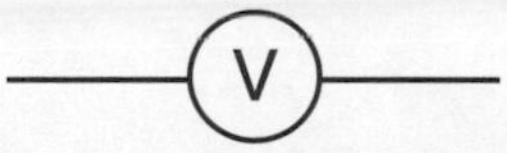 (V)	To measure voltage
Ammeter	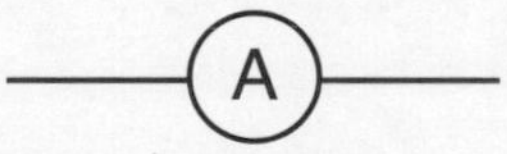 (A)	To measure current
Resistor		To cut down the amount of current flowing
Variable resistor		To control the amount of current flowing

Figure 16.1 Circuit symbols

> **Exam tip**
>
> A battery is two or more (electrical) cells. A battery can provide more electricity than a single cell. The battery in Figure 16.2 has two cells.

Circuit diagrams

Figure 16.2 shows a very simple electric circuit with one battery and one bulb.

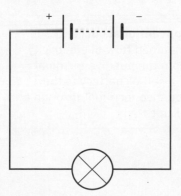

Figure 16.2 A simple electrical circuit

In an electrical circuit the wire must be a **conductor** (be able to conduct electricity). When the switch is open, the current flow stops as the circuit is not complete. (Air is an **insulator**; it will not conduct electricity.)

Voltage and current

Voltage is the amount of electrical energy supplied to a circuit (or a component). Batteries supply voltage; a voltage is measured using a **voltmeter** and the unit of voltage is the **volt** (**V**).

> **Voltage**: The amount of electrical energy supplied to a circuit; it is the voltage that causes an electrical current to flow.
>
> **Volt**: The unit of voltage (represented by the symbol V).

Current is the amount of electricity flowing around a circuit (or through a component).

Current is measured using an **ammeter** and the unit of current is the **ampere (amp)** (**A**).

Cells (and batteries) have positive and negative terminals (the + and – symbols, respectively). This is known as **cell polarity**.

> **Exam tip**
>
> If adding cells together (as in a battery) to provide a higher voltage, it is important that the negative terminal of one cell is connected to the positive terminal of the next cell.

Now test yourself

TESTED

1 What is the function of a voltmeter?
2 Define the term 'conductor'.
3 Name the unit of current.

Answers on p. 151

(H) In a circuit the current is carried by **electrons,** which are able to move. The electrons are repelled (expelled) from the negative terminal of a cell or battery and attracted towards the positive terminal. Understanding that an electrical current is a flow of electrons helps explain the difference between conductors and insulators; conductors have free electrons (that can move round the circuit) and insulators do not have free electrons.

Although an electric current actually flows from the negative terminal of a cell to the positive terminal, the direction of current (shown as arrows) in circuit diagrams (the **conventional current**) goes from positive to negative. This is because it was originally thought that electricity flowed from positive to negative and the tradition of showing current flowing this way remains.

Conductor: A material that allows electricity to flow through it easily.

Insulator: A material that does not allow electricity to flow through it easily.

> **Exam tip**
>
> Electricity will only flow if a circuit is complete, with no gaps.

Current: The amount of electricity flowing around a circuit (or through a component).

Ampere (amp): The unit of electrical current (represented by the symbol A).

Cell polarity: The existence of positive and negative terminals in a cell or battery.

Electrons: Negatively charged particles which can move around an electrical circuit conducting electricity.

Conventional current: The imagined flow of electricity from the positive terminal of a cell or battery to the negative terminal through a circuit.

Series and parallel circuits

Electric circuits can be arranged in series or in parallel.

Series circuits

In a **series circuit** the components (such as bulbs) are connected in sequence (one after another or side-by-side), as shown in Figure 16.3.

In the diagram, the switch is open – the circuit is incomplete – so the bulbs will not light. When the switch is closed, both bulbs will light.

In a series circuit:
- the **total voltage is shared** between the bulbs; in the circuit in Figure 16.3, each bulb receives half the voltage (assuming the two bulbs are identical)
- the **current does not change** (i.e. is the same for each component) around the circuit. This means that in the circuit in Figure 16.3, each bulb will have the same brightness (assuming they are identical).

> **Series circuit**: An electrical circuit in which the components are connected in sequence (one after another or side by side).

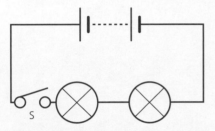

Figure 16.3 Two bulbs connected in series

> **Exam tip**
>
> If the circuit in Figure 16.3 had three bulbs, each would get one third of the voltage supplied by the battery.

Parallel circuits

In a **parallel circuit** there is more than one branch through which electricity can flow. Figure 16.4 shows two bulbs connected in parallel.

> **Parallel circuit**: A circuit which has more than one branch through which electricity can flow.

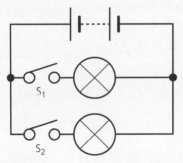

Figure 16.4 Two bulbs connected in parallel

In Figure 16.4, if switch S_1 is closed, only the upper bulb will light because current will flow through the upper branch of the circuit only. If both switches (S_1 and S_2) are closed, both bulbs will light. If only S_2 is closed then only the lower bulb will light.

In a parallel circuit:
- each bulb (or branch of the circuit) receives the **total voltage** from the battery
- the **current is shared** between the branches (i.e. the total current taken from the supply is the sum of the currents through the separate components).

Measuring voltage and current

When measuring the voltage *across* a component, such as a bulb, the voltmeter is connected **in parallel** with it.

When measuring the current *through* a component, such as a bulb, the ammeter is placed **in series** (see Figure 16.5).

Example

Figure 16.5 represents an electric circuit containing two cells and three identical bulbs.

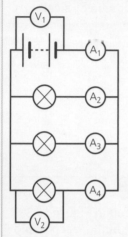

Figure 16.5 **A circuit with three bulbs**

(a) Describe how the bulbs are arranged.
(b) Ammeter A_1 has a reading of 6 A. What will be the reading on ammeter A_2?
(c) Voltmeter V_1 has a reading of 12 V. What will be the reading on voltmeter V_2?
(d) Give **one** change that will take place if more cells are added to the circuit.

Answer

(a) Parallel
(b) 2 A: the current is shared among the three branches in a parallel circuit (6 A/3 = 2 A)
(c) 12 V: each branch receives the total voltage in a parallel circuit
(d) Increased voltage/increased current/increased brightness of bulbs

Prescribed practical P1

Describe and carry out an experiment using a voltmeter to measure the voltage across a metal wire and an ammeter to measure the current passing through the wire

The data collected can be used to draw a *V–I* graph, where *V* is a measure of the voltage in volts and *I* is the current in amps. A *V–I* graph is a straight line passing through the origin showing that, for a metal wire at constant temperature, the current and voltage are proportional. This is known as **Ohm's law**.

> **Ohm's law**: When electricity flows through a metal wire at constant temperature, the voltage and current are directly proportional to each other.

Exam practice answers at **www.hoddereducation.co.uk/myrevisionnotesdownloads**

Now test yourself

TESTED

4 What is meant by the term 'series circuit'?
5 How is the voltage distributed between the bulbs in a series circuit?
6 How is the current distributed between the branches of a parallel circuit?

Answers on p. 151

Resistance

REVISED

Circuits and their components resist (oppose) the electric current flowing through them; this is described as **resistance**. The unit of resistance is the **ohm** (Ω).

Some facts about resistance:

● The bigger the resistance, the less current will flow.
● Insulators, such as plastic, have a high resistance that prevents electricity flowing through them.
● Conductors, such as metals, have low resistance.
● If you add more bulbs to a series circuit, the bulbs get dimmer. This is because each time a bulb is added, there is more resistance.

Voltage (V), current (I) and resistance (R) are linked by the equation:

$$V = I \times R \text{ (or } R = \frac{V}{I})$$

You can calculate resistance by measuring the voltage and the current in a circuit (Figure 16.6) and then using the formula:

$$R = \frac{V}{I}$$

Factors affecting the resistance of a wire

● **Length**: the longer a wire is, the more wire there is to resist the current. If the length of wire is doubled, the resistance is doubled and so on. The resistance of a wire is directly proportional to its length.
● **Temperature**: the resistance in a metal wire increases with temperature.

H ● **Material**: the lower the resistance a material has, the better it is at conducting. This is why copper is commonly used in circuits. Wires often used to demonstrate resistance are constantan (a mixture of nickel and copper) and nichrome (a mixture of nickel and chromium).
● **Cross-sectional area**: the thicker the wire is, the smaller is its resistance. If the cross-sectional area is doubled, the resistance will halve and so on. The resistance of a wire is inversely proportional to its cross-sectional area.

You should be able to describe how to investigate the idea that resistance of a metallic conductor at constant temperature depends on the area of cross-section and the material it is made from.

> **Resistance**: The opposition by a material to the flow of electrical current.
>
> **Ohm**: The unit of resistance (represented by the symbol Ω).

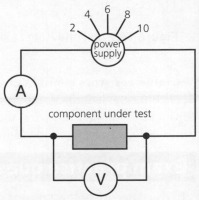

Figure 16.6 Measuring resistance using a voltmeter and an ammeter

Prescribed practical P2

Investigate experimentally how the resistance of a metallic conductor at constant temperature depends on length

Obtain a number of values for resistance (by using the formula for resistance when you have values for voltage and current) using different lengths of wire (Figure 16.7) and plot a graph of resistance (y-axis) against length (x-axis). The graph will be a straight line that passes through the origin, showing that when a metal wire is at constant temperature, the resistance and length of wire are proportional.

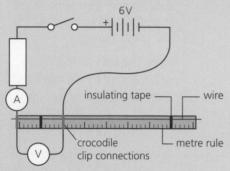

Figure 16.7 Investigating resistance in different lengths of wire

> **Exam tip**
>
> When planning investigations involving resistance, you should take precautions to make sure the metal wire stays at a constant temperature, for example, not taking replicates until the wire is at room temperature.

Variable resistors control current in a circuit by increasing or decreasing the length of resistance wire that the current has to flow through: the longer the wire, the more resistance. They are used in dimmer light switches and volume controls in televisions and radios.

Exam practice questions

1 The diagram below shows two electrical circuits.

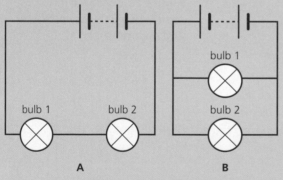

Figure 16.8 Two electrical circuits

    (a) In which circuit are the bulbs connected in parallel? [1]

    (b) If the batteries supply 6V, what will be the voltage across bulb 2 in each circuit? Explain your answer. [4]

    (c) What will happen in each circuit if bulb 2 becomes loose? [2]

2 A bulb in an electrical circuit has a voltage of 5V and a current of 1.25A. Calculate the resistance. [2]

3 (a) Describe how you could investigate the effect of the cross-sectional area of a wire on its resistance. [4]

    (b) State **two** things you would need to keep the same in this investigation. [2]

Answers online

ONLINE

17 Household electricity

Electricity is relatively easily distributed to homes and businesses. It can then be transferred into other forms of useful energy easily. These include:

- **heat** in irons, ovens and toasters
- **movement** (kinetic) in fans, washing machines and tumble dryer motors
- **sound** in radios, televisions and electric doorbells
- **light** in bulbs and TV screens.

It is important to remember that electricity is **dangerous**. All the necessary safety precautions must be taken when working with electricity.

Protection from electrical shock

REVISED

The design and wiring of a three-pin plug show many safety features.

Three-pin plugs

Figure 17.1 shows the wiring inside a three-pin plug.

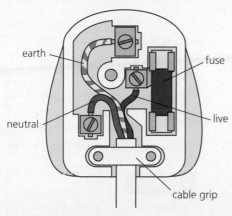

Figure 17.1 The wiring in a three-pin plug

Table 17.1 describes the colours and functions of the live, neutral and earth wires in a three-pin plug.

Table 17.1 The colours and functions of the live, neutral and earth wires

Colour	Name	Connection	Function
Brown	Live	Right terminal	Electricity enters the plug and travels to the appliance
Blue	Neutral	Left terminal	Returns the electricity to the plug socket from the appliance
Yellow/green	Earth	Top terminal	If a fault develops in an appliance that has metal parts, the metal can conduct electricity; the earth wire is an escape route allowing the electricity to flow to earth and not through someone touching the appliance

Safety features of the three-pin plug

These include the following:

- The casing of the plug is made of plastic, an insulator.
- The cable grip prevents the cable (and wires) from being pulled out.
- The fuse protects the appliance and the user.

Fuses

Fuses are short lengths of safety wire that stop too much current flowing in a circuit. They are very important **safety** devices.

This is how a fuse works:

1 The more current in the fuse wire, the hotter it becomes.
2 If the current is too high, the fuse wire melts ('blows').
3 The circuit is no longer complete (there is a gap).
4 This stops the flow of current and protects the appliance and the user.

Common fuse sizes are 1 A, 2 A, 3 A, 5 A and 13 A. The correct fuse must be used for each appliance.

- The fuse should not melt when the normal current is flowing through the circuit.
- The fuse should melt when the current rises just above the normal level.

The formula:

power (P) = voltage (V) × current (I)

can be used to calculate the current normally flowing through an appliance.

The units of **power** (the rate at which electrical energy is transferred) are **watts** (**W**) or **kilowatts** (**kW**) (1000 W = 1 kW).

$$P = V \times I$$

can be rearranged as

$$I = \frac{P}{V} \text{ or } V = \frac{P}{I}$$

Worked example

Find the fuse required for a 200 W lamp working at a mains voltage of 230 V.

Answer

Power = 200 W and voltage = 230 V.

$$I = \frac{P}{V}$$

$$= \frac{200}{230} = 0.87 \text{ A}$$

So it is best to use a 1 A fuse; it is just bigger than the current normally needed by the appliance.

Exam tip

It is important that appliances with metal cases are earthed as metal can conduct electricity.

Exam tip

The three wires in the plug are insulated with a plastic coating. However, if you are asked to name a safety feature in a plug, you may not gain credit for this as this safety feature is not actually part of the plug.

Fuse: A safety device consisting of a fine wire which melts if too much current flows through it, thus breaking an electrical circuit.

Power: The rate at which electrical energy is transferred or work is done.

Watt (kilowatt): The unit of power, represented by the symbol W (kW).

Exam tip

When using the formula, power must be given in watts. If you are given power as kilowatts you must convert this to watts (×1000) before using the formula.

Exam tip

If a fuse with too low a value is used, it will 'blow' when the appliance is turned on, even when there is not a problem. If a fuse with too high a value is used, it will allow current to flow even when there is a problem with the appliance, putting the appliance and the user at risk.

Exam tip

The fuse is on the live side of the plug. If the fuse blows, this will stop electricity reaching the appliance and the user.

Other electrical safety features

Another safety feature of many electrical appliances is **double insulation** in which the appliance, such as a computer, has a plastic casing that cannot give an electric shock, even if there is a fault inside. These appliances have their metal parts in a separate plastic container within the outer plastic cover; they are **double insulated**. Such appliances do not need a plug with an earth wire.

> **Double insulation**: A safety system which encloses all conducting parts of an electrical circuit in a plastic box, so that the user can never touch a live component and get an electrical shock.

Now test yourself

TESTED ☐

1 State the colour of the live wire in a three-pin plug.
2 State the function of the neutral wire.
3 Name the unit of power.

Answers on p. 151

The cost of electricity

REVISED ☐

From information provided, you need to be able to calculate how much electricity is used over a period of time and how much it costs.

Electricity meters are used to show how much electricity a house or business uses. Companies that supply electricity refer to '**units**' of electricity. Each unit costs a certain amount, so:

electricity bill = number of units used × cost of each unit

H The unit of electricity is the **kilowatt-hour** (**kWh**). This is the amount of electricity used by an appliance that uses 1000 W of power for 1 hour.

The cost of electricity used by an appliance can be calculated using the formula:

energy (kWh) = power (kW) × time (hr)

> **Exam tip**
>
> Many examination questions show the meter readings at the start and at the end of a billing period. You can work out the cost by calculating the difference in the readings (the number of units used) and multiplying this by the cost of each unit.

> **Kilowatt-hour**: A 'unit' of electricity, represented by the symbol kWh. It is the amount of electricity used by an appliance that uses 1000 W of power for 1 hour.

Worked examples

Calculate the cost of using a cooker ring with a power of 2000 W for 2 hours if the cost of electricity is 10p per kWh.

Answer

power = 2000 W = 2 kW; time = 2 hr

energy used = 2 × 2

     = 4 kWh

cost = 4 × 10p = 40p

> **Exam tip**
>
> In this formula the power is in kW. In questions about the cost of electricity used, many students forget to convert information given in watts to kilowatts.

Now test yourself

TESTED ☐

4 In terms of electricity usage, explain the term 'kilowatt-hour'.
5 State the symbol for kilowatt-hour.

Answers on p. 151

Saving electricity

You can reduce your electricity bills by using less electricity. This can be done by:

- turning off electrical equipment when it is not in use, for example not leaving the television on 'stand-by'
- using low-energy light bulbs
- turning off lights when a room is not being used
- buying energy-efficient appliances.

Exam practice

1 (a) Describe **three** safety features of a three-pin plug. [3]
  (b) Suggest why an earth wire has a low resistance. [1]
2 (a) Explain how a fuse works as a safety device. [3]
  (b) Explain fully why it is important to use a fuse with the correct rating. [2]
3 Table 17.2 shows how the number of cells in a battery affects voltage and current.

Table 17.2

Number of cells	Voltage/V	Current/A
1	2	0.2
2	4	0.3
3	6	0.4
4	8	0.5

  (a) Give **one** trend shown by these results. [1]
  (b) Using the equation power = voltage × current, calculate the power used in watts when the battery has four cells. [1]
4 Calculate the cost of using an electric fire with power 3500 W for 1 day if the cost of electricity is 15p per kWh. [3]

Answers online

ONLINE

18 Energy

Energy comes in many different forms. The main types of energy are summarised in Table 18.1.

Table 18.1 Different types of energy

Type of energy	Description
Chemical	Energy stored in coal, oil, food, batteries
Heat	Energy released by a fire
Electrical	The energy provided by electrical current; electrical energy is used to drive electrical appliances
Sound	Vibrations in the air; the type of energy that allows us to hear
Light	Energy provided by the Sun and light-emitting devices such as televisions
Magnetic	The energy around a magnet
Strain energy	The energy stored in an object when compressed or stretched, e.g. an elastic band
Kinetic	Energy due to movement
Gravitational potential	Energy due to position in terms of the Earth's gravitational field (height above the ground); this type of energy causes water to flow downhill

Energy transfers

REVISED

You should be able to work out and show energy transfers. For example,

- microphone: sound energy → electrical energy
- loudspeaker: electrical energy → sound energy
- coal burning in a fire: chemical energy → heat + light energy.

The unit of energy is the **joule** (**J**). Large quantities can be measured in **kilojoules** (**kJ**). There are 1000 joules in a kilojoule.

> **Joule**: The unit of energy, represented by the symbol J.

Conservation of energy

It is possible to transfer energy from one type to another. For example, when we burn coal, some of the chemical energy in the coal is transferred to heat and light energy. The reduction in chemical energy in the coal will be the same as the sum of the heat and light energy released. This will apply to any energy transfer: the amount of energy at the end of the process will be the same as the amount of energy at the start.

This is known as the **Principle of the Conservation of Energy**: energy can be changed from one form to another but the total amount of energy does not change.

> **Principle of the Conservation of Energy**: Energy can be changed from one form to another but the total amount of energy does not change.

Worked example

A battery is connected to a lamp in a circuit. In lighting the bulb, the battery's store of chemical energy falls by 100 J. The lamp releases 80 J of light energy. How much heat energy was released by the lamp?

Answer

100 = 100 − 80

  = 20 J

Using formulae to calculate energy transfers

Kinetic energy

Kinetic energy is the energy possessed by a moving object; stationary objects do not have kinetic energy.

Kinetic energy (E_k) can be calculated using the equation:

$$E_k = \tfrac{1}{2}mv^2$$

where E_k is in joules (J); mass (m) is in kilograms (kg); and the velocity (speed) (v) of an object is in metres per second (m/s).

> **Kinetic energy**: The energy possessed by an object due to its movement.

Worked example

The kinetic energy in a moving car is 32 400 J. It is travelling at a speed of 6 m/s. Calculate the mass of the car.

Answer

$$E_k = \tfrac{1}{2}mv^2$$

$$32\,400 = \tfrac{1}{2} \times m \times (6)^2$$

$$m = \frac{2 \times 32\,400}{36}$$

$$m = 1800\,\text{kg}$$

> **Exam tip**
>
> Remember that Higher-Tier students have to be able to rearrange formulae in calculations.

Gravitational potential energy

Gravitational potential energy is the energy an object has as a result of its position (height) above the ground.

The gravitational potential energy will be transferred into kinetic energy if the object falls toward Earth due to gravity.

Gravitational potential energy (E_p) can be calculated using the formula:

$$E_p = mgh$$

$$\text{gravitational potential energy} = \text{mass (kg)} \times \text{gravitational field strength (N/kg)} \times \text{height (m)}$$

For all calculations that you will meet in your exam, g (the gravitational field strength) = 10 N/kg.

> **Gravitational potential energy**: The energy possessed by an object because of its height above the ground.

> **Gravitational field strength**: A measure of how strong the force of gravity is. On Earth, it is about 10 N/kg.

Exam practice answers at **www.hoddereducation.co.uk/myrevisionnotesdownloads**

Worked example

An object has a mass of 40 kg. It is raised by a pulley from a height of 1 metre to 5 metres above ground. Calculate the increase in gravitational potential energy.

Answer

Initial:

$E_p = mgh$

$= 40 \times 10 \times 1 = 400\,J$

Final:

$E_p = mgh$

$= 40 \times 10 \times 5 = 2000\,J$

increase in $E_p = 1600\,J$

Efficiency

Not all the energy used in a particular process or device is useful. For example, the electrical energy supplied to a television can be transferred into visible (light), sound and heat energy. The visible and sound energy is useful but the heat energy is waste energy.

The **efficiency** of an appliance is a measure of how good it is at transferring useful energy from one form to another.

The efficiency is calculated using the equation:

$$\text{efficiency} = \frac{\text{useful output energy}}{\text{total input energy}}$$

with efficiency shown as a decimal or a percentage.

> **Efficiency**: A measure of how much of the input energy (to a process or device) is converted to useful output energy.

> **Exam tip**
>
> Efficiency can be shown as a percentage or as a decimal. An efficient device can approach 100% or 1.0 but can never exceed these values.

Exam practice

1 Copy and complete the sentence below.
The Principle of the Conservation of Energy states that energy can be _____ from one form to another but the _____ amount of energy does not change. [2]

H 2 A conveyor belt lifts a 100 kg box of bricks from the ground to a height of 6.5 metres in a building. Calculate the increase in gravitational potential energy in joules. [3]

3 Table 18.2 gives information about two types of light bulb.

Table 18.2

	Standard filament bulb	Low-energy bulb
Electrical power input	40 W	12 W
Light power output	4 W	4 W
Efficiency/%	0.10	

(a) Calculate the efficiency of the low-energy bulb using the equation: [2]

$$\text{efficiency} = \frac{\text{useful output energy}}{\text{total input energy}}$$

(b) Using the information provided, explain fully **one** reason for using low-energy bulbs. [2]

(c) Suggest **one** way a filament bulb wastes energy. [1]

Answers online

ONLINE

19 Electricity generation

Making electricity

Moving a **magnet** inside a coil of wire makes electricity and generates a current in the wire.

Figure 19.1 shows a simple **dynamo**; the rotation of the magnet produces electricity in the coils of wire.

> **Dynamo**: A device for converting the kinetic energy of a moving object into electricity.

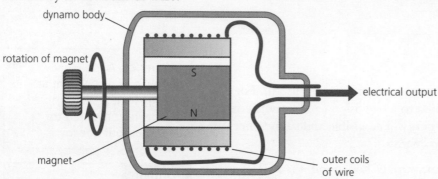

dynamo body

rotation of magnet

S

N

electrical output

magnet

outer coils of wire

Figure 19.1 Generating electricity using a dynamo

The same principle is used in generating electricity in power stations. The key difference is that power stations make much more electricity.

Power stations

Figure 19.2 shows the main parts of a power station.

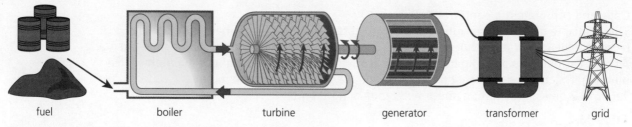

fuel   boiler   turbine   generator   transformer   grid

Figure 19.2 A fossil fuel power station

The functions of the different parts of a power station are summarised in Table 19.1.

Table 19.1 The functions of, and energy changes in, the component parts of a power station

Component	Function	Energy change
Burner	The fuel is burned to produce heat	Chemical energy in fossil fuels is transferred to heat (thermal) energy
Boiler	Heat is used to turn water into steam	Heat energy is used to boil water, forming steam (which has kinetic energy)
Turbine	The steam drives the **turbine**, which is connected to the **generator**	The kinetic energy of the steam is transferred to the kinetic energy of the turbine (which in turn drives the generator)
Generator	The rotating generator (large dynamo) makes the electricity	Kinetic energy is transferred to electrical energy

Turbine: A machine (in a power station) that drives the generator (converts the kinetic energy of the turbine into the kinetic energy of the generator). The blades of the turbine turn due to the kinetic energy of the steam.

Generator: The part of a power station that makes electricity (similar in principle to a dynamo) by converting kinetic energy to electrical energy.

ⓗ Electricity transmission

REVISED

Electricity has to be transmitted from where it is made (power stations) to houses and businesses around the country. The electricity is distributed through a (national) **grid** of power lines.

The electricity being transmitted in the (national) grid:
- is transmitted at high voltage
- has a relatively low current
- is often carried in power lines suspended from high pylons for safety because the high voltages used are very dangerous.

The high voltages used allow the electricity to flow with a relatively low current and less energy is wasted as heat. It also allows thinner (and cheaper) power lines to be used.

Transformers

A typical power station generates electricity around 25–30 kV. This is then converted to the grid voltage (around 400 kV) using a **step–up transformer**. Before the electricity can be used in homes or businesses, the voltage is reduced using a **step–down transformer**.
- Step-up transformer: increases the voltage and decreases the current (to reduce energy (heat) losses).
- Step-down transformer: decreases the voltage and increases the current.

Figure 19.3 shows the position of step-up and step-down transformers in the grid system.

> **Transformer**: A device which converts high voltages to low voltages and vice versa.

> **Exam tip**
>
> Remember that a step-up transformer increases voltage and decreases current and a step-down transformer decreases voltage and increases current.

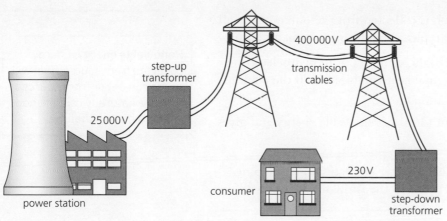

Figure 19.3 The position of step-up and step-down transformers in an electricity grid

Once the electricity reaches homes it is used in many ways. It can be transferred into heat energy (in electric kettles and toasters), kinetic energy (in motors in food mixers), light energy (in light bulbs and televisions) and sound energy (in radios and televisions). This ability to be transferred into so many energy forms makes it a very useful energy type.

Although electricity is very useful, we must not forget that most electricity in the British Isles is made from the burning of fossil fuels (a non-renewable source of fuel).

Now test yourself

TESTED

1 Name the part of a power station that turns (drives) the generator.
2 Describe the energy changes that take place in the generator.
3 In terms of voltage and current, describe the changes that take place in a step-up transformer.

Answers on p. 152

Renewable and non-renewable sources of energy

REVISED

Non-renewable sources

A **non-renewable energy** source is one that has a finite supply (it will run out at some time).

Fossil fuels (such as coal, oil and natural gas) are considered as non-renewable as their formation takes millions of years. Fossil fuels are formed from the remains of dead plants and animals subjected to high temperatures and pressures over millions of years in the Earth's crust.

Nuclear energy, based on nuclear fission, is also non-renewable as it relies on supplies of uranium ore which will also run out over time.

Biomass (energy from the combustion of wood) is non-renewable if it uses plant material that is not regrown. However, the use of biomass from willow can be renewable if it is allowed to regrow.

Renewable sources

Renewable energy is energy which is collected from resources that will never run out or that are naturally replenished during a human lifetime.

Renewable energy sources include light, wind, rain (in hydroelectric plants), tidal, waves, wood, geothermal heat and biomass (if the plant material is regrown).

Table 19.2 summarises the different forms of renewable energy and gives advantages and disadvantages.

In recent years there has been a much greater emphasis on developing renewable sources of energy. This is due to the finite nature of fossil fuels.

> **Non-renewable energy**: Energy from a source which has a finite supply (it will eventually run out).

> **Exam tip**
>
> When learning about fossil fuel formation, think of three key points:
>
> **What from?** Dead plants and animals
>
> **How?** High temperatures and pressures
>
> **How long?** Millions of years

> **Renewable energy**: Energy collected from sources that will never run out or from sources that are naturally replenished within a human lifetime.

> **Exam tip**
>
> Fossil fuels produce carbon dioxide when burnt. Carbon dioxide is a 'greenhouse' gas and is a major contributor to global warming. This is another very important reason why there is an increasing emphasis on renewable energy sources as opposed to the combustion of fossil fuels. However, greenhouse gases and global warming are covered in the chemistry section of this course and therefore will not be expected as answers in Unit 3 (Physics) papers.

Table 19.2 Renewable energy sources

Renewable energy source	Method	Advantages	Disadvantages
Wind	Blades of wind turbine rotate and drive a generator	● Cost effective once built ● Unlimited supplies of wind	● Only works when wind blows ● Noisy and unsightly
Rain (hydroelectric)	Rotating turbine drives a generator	● Cost effective once built ● Unlimited supplies of water	● Land is flooded to create a reservoir, ruining habitats and destroying farmland ● Very expensive to build
Tidal	Tides coming in and out drive turbines linked to generators	● Cost effective once built ● Unlimited supplies of water	● Destroys habitats ● Very expensive to build
Wave	Force of waves drive turbines linked to generators	● Cost effective once built ● Unlimited supplies of wave energy	● Expensive ● Relatively inefficient
Light (solar)	Solar cells use energy directly from the Sun to generate electricity	● Cost effective once built ● Unlimited supplies of light energy	● Expensive to install ● Can be unreliable due to weather
Wood (biomass)	Trees, e.g. willow, grown and harvested over a short-term cycle	● Unlimited supplies of wood ● No expensive equipment required	● Labour intensive
Geothermal	Hot water (or steam) from deep underground can be piped up to drive turbines linked to generators	● Unlimited supplies ● Can also provide hot water directly to houses (avoiding heating costs)	● Only suitable in some countries

Worked example

In Northern Ireland, willow is grown and used as a biofuel to provide heat energy. Willow can be grown in 3–4 year cycles between harvesting.

(a) Why can willow be classed as a renewable energy source?

(b) Give **one** advantage and **one** disadvantage of using wood (biofuel) as an energy source compared with solar power.

Answer

(a) The willow is regrown (and so overall stock of wood is not decreased due to the burning of willow).

(b) Advantage: very little initial cost
Disadvantage: labour intensive

Now test yourself

TESTED ☐

4 Explain why nuclear energy is described as being non-renewable.
5 Give **one** disadvantage of tidal energy.

Answers on p. 152

Exam practice questions

1 (a) Describe how electricity is generated. [2]
  (b) Describe the role of a turbine in a fossil fuel power station. [2]
2 (a) In Figure 19.4 what is represented by **X** and **Y**? [2]

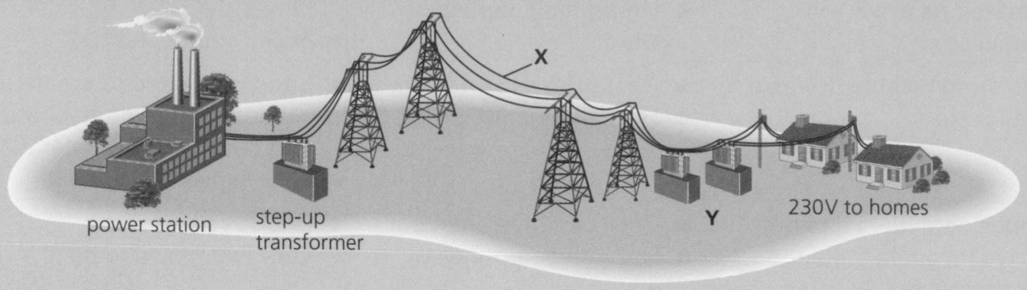

X

power station   step-up transformer   Y   230V to homes

Figure 19.4

  (b) Describe and explain the function of a step-up transformer. [3]
3 (a) Describe how fossil fuels are formed. [3]
  (b) Explain why there is an emphasis on replacing fossil fuels with renewable energy sources. [1]
  (c) (i) Name **one** example of a renewable energy source. [1]
    (ii) Give **one** advantage and **one** disadvantage of this energy source. [2]

Answers online

ONLINE

20 Heat transfer

Methods of heat transfer

REVISED

Heat transfer takes place by three methods:
- conduction
- convection
- radiation.

In all types of heat transfer it is heat (thermal) energy that is transferred.

Conduction

Most metals are good **conductors** of heat; this means they can transfer heat by the process of **conduction**. In metals the atoms are arranged very close together in a regular lattice (or pattern). Heat energy can pass rapidly across most metals.

Figure 20.1 shows how the ability of different materials to conduct heat can be compared.

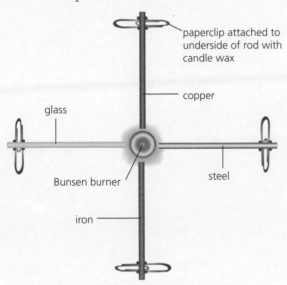

Figure 20.1 Investigating conduction

The paperclips will fall from the three metal rods long before the paperclip on the glass rod falls. This is because the metals conduct heat much better than glass. The heat will pass from the edge of the rods closest to the Bunsen burner along the metal rods by conduction, causing the candle wax to melt and the paperclips to fall. This shows that the metals are good conductors of heat. The glass is described as an **insulator** as it does not transfer (conduct) heat easily.

Conduction is important in everyday life:
- We use saucepans made of metal to allow the heat to be transferred from the hob to the food.
- We wear thick oven gloves to act as insulators to prevent the heat from hot pots burning our hands.

> **Conduction**: A process that describes the transfer of heat (thermal) energy through the particles of a solid. Materials, such as metals, that conduct heat easily are said to be good **conductors** of heat energy.

> **Insulator**: A material that does not transfer heat energy easily.

> **Exam tip**
>
> You should be able to work out the controlled variables (factors that should be kept the same) in this type of experiment. In an experiment similar to that shown in Figure 20.1:
> - the paperclips should be the same distance along the rods
> - the rods should be the same thickness
> - the same amount of candle wax is used to hold the clips in place
> - the ends of the rods closest to the Bunsen should be the same distance from the Bunsen.

Prescribed practical P3

Compare the heat conductivity of different materials by measuring the time it takes heat to travel through a variety of conductors and at least one insulator

Convection

Convection takes place in **liquids** and **gases**. If we heat a beaker of water, the water at the bottom of the beaker heats up and rises. This is seen by adding a dye to the bottom of a beaker of water.

As the water in Figure 20.2 is heated, the dye in the water moves up and through the water. The movement of the water is known as a **convection current**.

Convection: The transfer of heat energy in a liquid or a gas.

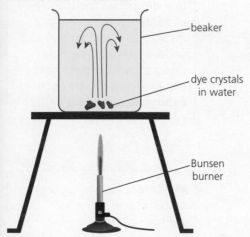

beaker

dye crystals in water

Bunsen burner

Figure 20.2 Demonstrating convection in liquids

A good example of convection in a gas is smoke rising from a hot fire or from a candle that has recently been extinguished.

In both liquids and gases the hot water or air rises and is replaced by cold water or gas, as shown in Figure 20.3. In both these examples of convection in the home, this pattern of air and liquid rising and falling creates continuous convection currents.

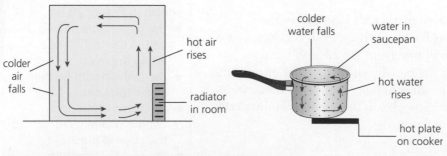

colder air falls

hot air rises

radiator in room

colder water falls

water in saucepan

hot water rises

hot plate on cooker

Figure 20.3 Examples of convection in the home

Hot air is less dense than cold air; this explains why the hot air rises and the colder air falls to replace the rising hot air. The same principle applies to hot liquids.

Exam practice answers at **www.hoddereducation.co.uk/myrevisionnotesdownloads**

Radiation

Radiation describes heat transfer that can pass through air (and other gases as well as a vacuum). Heat energy radiates out from hot objects. Radiation describes how the heat energy from the Sun travels through space to reach the Earth.

When there is a fire in a room, the hottest place will be just in front of the fire; this is because heat energy radiates out through the air.

> **Radiation:** The transfer of heat energy by waves passing out from a hot surface.

Now test yourself

TESTED

1 What is an insulator?
2 Name the process by which heat energy is normally transferred in liquids.
3 Name the process by which heat energy is transferred from the Sun to the Earth.

Answers on p. 152

Worked example

You should be able to describe how a fire in a room will heat the room.

Answer

A fire in a room heats the room by both convection and radiation, as shown in Figure 20.4.

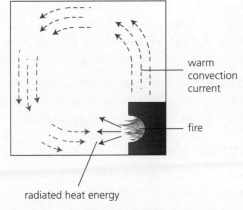

warm convection current

fire

radiated heat energy

Figure 20.4 **Heat transfer in a room**

Absorbing and radiating heat energy

Some surfaces are better than others at absorbing and radiating heat energy. Dark, matt surfaces are better at both absorbing and radiating heat energy than shiny surfaces.

If apparatus is set up as shown in Figure 20.5, the temperature sensor on the dark, matt surface will record a higher temperature than that on the shiny surface, showing that dark, matt surfaces absorb more radiant energy than light, shiny surfaces.

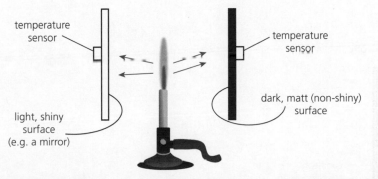

temperature sensor

temperature sensor

light, shiny surface (e.g. a mirror)

dark, matt (non-shiny) surface

Figure 20.5 Dark, matt surfaces are better at **absorbing** heat energy than light, shiny surfaces

Dark, matt surfaces are also better at radiating (emitting) heat energy than light, shiny surfaces, as can be seen in demonstrations similar to that shown in Figure 20.6.

In this investigation a metal box with a white, shiny surface and a dark, matt surface on different sides is filled with very hot water. Temperature sensors are placed a short distance away from each surface. The sensor close to the dark, matt surface will record a higher temperature than the sensor close to the white, shiny surface, showing that dark, matt surfaces radiate more heat than white, shiny surfaces.

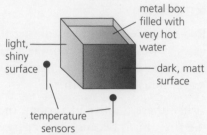

Figure 20.6 Dark, matt surfaces are also better at radiating heat energy than light, shiny surfaces

Heat transfer at the molecular level

REVISED

Conduction

The atoms in a metal are arranged in a regular structure and they are very close together. In metal atoms, the outer electrons are not tightly attached; they are **free** or '**delocalised**'. This means that these electrons can move easily throughout the whole metallic structure. As these electrons move, they carry heat energy from one part of a metal to another.

In insulators, the outer electrons are not free (or many fewer are free) so they cannot pass throughout the structure. The only way that heat energy can transfer in solid insulators is by passing from atom to atom – a much slower process.

> **Exam tip**
>
> Remember that, in metals, heat transfer is due to the electrons moving; the whole structure itself doesn't move.

Convection

When a liquid absorbs heat energy, the particles (atoms or molecules) gain energy and move further apart; they have more kinetic energy. As the particles are more widely spaced, they have a lower density than those in the cooler liquid. This lower density in the warmer liquid causes it to rise and it is replaced by cooler liquid of a higher density. This is the basis of convection currents. The same principle applies to convection in gases.

> **Exam tip**
>
> In heat transfer in liquids and gases by convection, the particles *do* move, transferring heat energy in the process.

Worked example

Figure 20.7 represents a demonstration that shows smoke going downwards. One end of a taper or something similar (e.g. tightly rolled paper) is lit and then the flame extinguished. The smoky end is held close to one of two glass chimneys in a container as shown. A lit candle is placed under the second chimney. Explain why the smoke coming from the smoky taper passes downwards into the container.

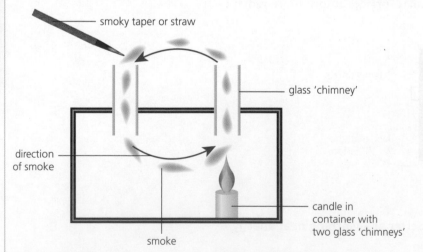

Figure 20.7 **Convection causing smoke to pass downwards**

Exam practice answers at **www.hoddereducation.co.uk/myrevisionnotesdownloads**

Answer

The candle heats up the air immediately above it causing it to become less dense. This air then rises up through the chimney and is replaced by cooler, more dense air coming down through the left chimney. This draws air (and smoke) down from the smoky taper. In summary, the smoke moves downwards as it is part of the convection current created by the candle.

Now test yourself

TESTED

4 Name the subatomic particles which transfer heat energy in metal conductors.
5 Explain, in terms of gas particles, why hot air rises.

Answers on p. 152

Conserving heat in the home

REVISED

Heat energy is lost in the home mainly through conduction and convection (through the walls, windows, roof and floor).

Heat loss can be reduced by:
- **cavity wall insulation** – two layers of brick with an insulator between each layer (Figure 20.8)
- **loft insulation** – a thick layer of insulating material used to floor the loft (Figure 20.9)

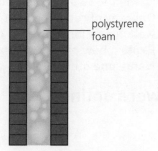

Figure 20.8 Cavity wall insulation reduces heat loss

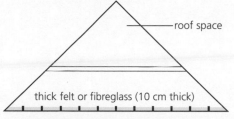

Figure 20.9 Loft insulation

- **double glazing** – two layers of glass separated by air (an insulator) (Figure 20.10)
- **carpets** – having a carpet will reduce heat loss through the floor.

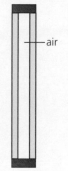

Figure 20.10 Double glazing

Exam tip

In many modern houses, loft insulation is the most effective method of reducing heat loss.

Exam practice questions

1 Figure 20.11 represents a saucepan.

Figure 20.11

Explain why this saucepan is made of three different materials. [3]

2 Figure 20.12 shows two thermometers. In one, the bulb has been painted black.

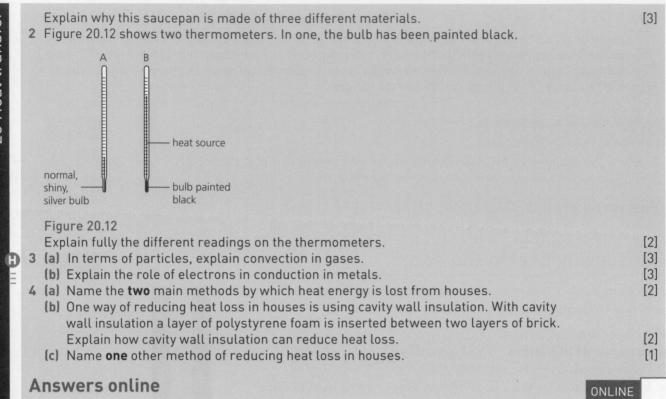

normal, shiny, silver bulb

heat source

bulb painted black

Figure 20.12

Explain fully the different readings on the thermometers. [2]

3 (a) In terms of particles, explain convection in gases. [3]

(b) Explain the role of electrons in conduction in metals. [3]

4 (a) Name the **two** main methods by which heat energy is lost from houses. [2]

(b) One way of reducing heat loss in houses is using cavity wall insulation. With cavity wall insulation a layer of polystyrene foam is inserted between two layers of brick. Explain how cavity wall insulation can reduce heat loss. [2]

(c) Name **one** other method of reducing heat loss in houses. [1]

Answers online

ONLINE

21 Waves

Waves are **vibrations** that carry **energy** from one place to another.

Types of waves

There are two types of wave motion.

1 **Longitudinal waves**: the particles vibrate in the **same direction** (are parallel) to the direction of travel. **Sound** and ultrasound are examples of longitudinal waves.

In the slinky spring in Figure 21.1, the particles vibrate in the same direction (plane) that the wave is travelling (along the spring).

> **Longitudinal wave**: A wave in which the particles vibrate in the same direction as (parallel to) the direction of travel.

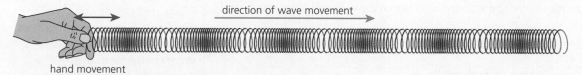

direction of wave movement

hand movement

Figure 21.1 A longitudinal wave

2 **Transverse waves**: the particles vibrate at right angles to the direction that the wave is travelling. **Electromagnetic** waves (including light) and water waves are examples of transverse waves.

In the slinky spring in Figure 21.2, the particles vibrate at right angles (are perpendicular) to the direction in which the wave is travelling.

> **Transverse wave**: A wave in which the particles vibrate at right angles to the direction of wave travel.

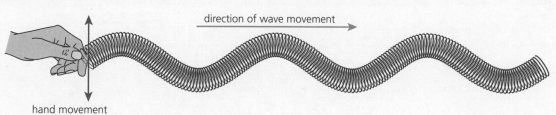

direction of wave movement

hand movement

Figure 21.2 A transverse wave

Features of transverse waves

Waves have three important features: wavelength, frequency and amplitude.

- The **wavelength** is the distance between two successive crests or troughs.
- The **amplitude** is the maximum height of a wave.
- The **frequency** is the number of waves passing a particular point in one second.

> **Wavelength**: The distance between two successive wave crests or troughs.
>
> **Amplitude**: The maximum height of a wave
>
> **Frequency**: The number of waves passing a particular point in one second.

Figure 21.3 shows the wavelength and amplitude of a typical transverse wave.

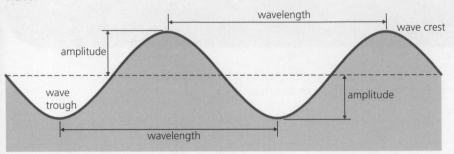

Figure 21.3 Wavelength and amplitude

Frequency is measured in **hertz** (**Hz**). A wave that has one complete wave passing a particular point each second has a frequency of 1 Hz.

Hertz: The unit of frequency, represented by the symbol Hz.

The wave equation

The wave equation gives the relationship between velocity, frequency and wavelength:

$$v = f\lambda$$

velocity = frequency × wavelength

with the velocity of the wave in metres per second (m/s), frequency in hertz (Hz) and the wavelength in metres (m).

Exam tip

Diagrams similar to Figure 21.3 appear in exam papers often. Many students incorrectly measure amplitude as the distance from the wave crest to the wave trough – they double the correct value.

Worked example

Calculate the velocity of a wave that has a wavelength of 5 cm and a frequency of 2 Hz.

Answer

$v = f\lambda$

$v = 2 \times 0.05$

$v = 0.1$ m/s

Sound

REVISED

Sound has the following features:
- It is a longitudinal wave.
- It is produced by vibrating objects that cause the air to vibrate.
- It cannot travel through a vacuum.
- Bigger vibrations → bigger amplitudes → louder sounds.

An **echo** is a reflected sound. Hard surfaces **reflect** most of the sound that hits them; soft surfaces **absorb** most of the sound and reflect less.

Echo: Reflected sound.

Echoes can be useful, but sometimes they are not. Echoes can create problems in concert halls and reduce the quality of music. To reduce echoes, many concert halls (auditoria) have soft furnishings, such as curtains and carpets, covering the walls and floors. These furnishings absorb rather than reflect sound.

Measuring the speed of sound

The speed of sound can be measured in different ways.

The echo method

1 Two students stand together, some distance away from a hard structure like a wall or building.
2 One student bangs two wooden blocks together.
3 The other student uses a stopwatch to measure the time between hearing the original bang and hearing its echo.
4 The speed of sound can be calculated using the equation:

$$\text{speed} = \frac{\text{distance}}{\text{time}}$$

Worked example

Two students carried out an investigation to measure the speed of sound using the echo method described above. The wall was 290 metres away and the average time between the initial sound and the echo was 2 seconds.

(a) Calculate the speed of sound using the formula

$$\text{speed} = \frac{\text{distance}}{\text{time}}$$

(b) The speed of sound in air is normally 330 m/s. Suggest one reason why the students' value was slightly different from this.

Answer

(a) $\text{speed} = \dfrac{580\,\text{m (distance to the wall and back)}}{2\ \text{seconds}} = 290\,\text{m/s}$

(b) Slow reactions of the time keeper.

⊕ The flash-bang method

This method does not involve echoes; it involves two people who are a considerable distance apart, for example 1 kilometre. One person is in a car and the other (the recorder) is a long distance away with a stopwatch. They must be able to see each other.

1 The person in the car flashes the headlights and sounds the horn at the same time.
2 Immediately the lights are seen by the other person, the stopwatch is started.

H 3 Once the sound is heard, the stopwatch is stopped.
4 The person in the car and the recorder change positions (to cancel out wind effects) and the process is repeated.
5 Several readings can be taken in each position to improve reliability.
6 The speed of sound is calculated using the formula:

$$\text{speed} = \frac{\text{distance}}{\text{time}}$$

Figure 21.4 The flash–bang method of measuring the speed of sound

Hearing and ultrasound

Hearing

The **audible hearing range** covers the frequencies that humans can hear. The audible hearing range is from **20 Hz to 20 kHz**. Many older people find it difficult to hear sounds over a frequency of around 14 kHz.

Ultrasound

Ultrasound has a frequency **higher than 20 kHz**. Ultrasound cannot be heard by humans.

Ultrasound has many uses including:
- **foetal scanning** – checking the health and size of a foetus in the womb
- checking for problems in **body organs** – for example looking for gallstones
- **measuring distances** – ultrasound scanners are used by estate agents to quickly and accurately measure room sizes; boats use ultrasound to calculate the depth of water or to find shoals of fish; parking sensors in modern cars can use ultrasound
- identifying **microscopic cracks** in structures such as railway lines or large buildings.

Sonar (a particular technique using ultrasound or audible sound) is a technique that can be used by submarines to detect other shipping or by fishing boats to detect fish. Pulses of sound can be emitted and the presence of echoes can detect other objects. The actual distance these objects are away can be calculated using the equation:

distance = speed of sound × time

> **Exam tip**
>
> When describing the flash–bang method, it is important to state that the car and the person with the stopwatch are a large distance apart– at least 1 km. Otherwise, the time difference between seeing the lights and hearing the horn is too short to measure accurately.

> **Audible hearing range**: The range of frequencies that humans can hear (20 Hz to 20 kHz).

> **Ultrasound**: Sound waves with a frequency higher than 20 000 Hz (20 kHz).

Now test yourself

TESTED

4 What is the audible hearing range in humans?
5 Define the term 'ultrasound'.

Answers on p. 152

The electromagnetic spectrum

The **electromagnetic spectrum** is the range of electromagnetic waves. They range from very shortwave gamma radiation (about 0.000 000 000 01 m) to longwave radio waves (about 1000 m).

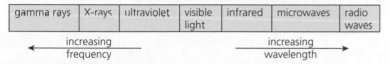

Figure 21.5 The electromagnetic spectrum

> **Electromagnetic spectrum**: The range of electromagnetic waves in order of increasing wavelength and decreasing frequency (gamma rays – X-rays – ultraviolet – visible light – infrared – microwaves – radio waves)

All electromagnetic waves:
- carry energy
- travel as transverse waves
- travel at the same speed through a vacuum.

Uses of electromagnetic waves

Table 21.1 details some of the uses of different waves in the electromagnetic spectrum.

Table 21.1 Some uses of electromagnetic waves

Electromagnetic wave	Use
Radio waves	<ul><li>Television and radio</li><li>Radar uses radio waves to track boats and planes</li></ul>
Microwaves	<ul><li>Satellite television</li><li>Mobile phones</li></ul>
Infrared	<ul><li>TV remote controls and heating from electrical heaters</li><li>Infrared cameras can take photographs at night</li></ul>
Visible light	<ul><li>Telephone networks using fibre optic communication systems</li><li>Lasers that emit high-energy visible light can be used during surgery</li></ul>
Ultraviolet	<ul><li>Some substances can absorb energy from UV light and then emit this energy as visible light (fluorescence)</li><li>Fluorescence can be used in security systems and in developing fingerprints</li></ul>
X-rays	<ul><li>X-rays can pass through flesh but not more dense objects, such as bone, and so can be used in hospitals to identify broken bones or obstructions in the body</li><li>Dentists use X-rays to check the health of teeth</li></ul>
Gamma rays	<ul><li>Gamma radiation is used in radiotherapy to kill cancer cells</li><li>Gamma rays can also be used to sterilise medical equipment</li></ul>

Mobile phones

Mobile phones send and receive messages by **microwaves**. The signal from a mobile phone is sent to the nearest communications mast. From there the signal is sent through a series of masts that act as **repeater stations** until the signal reaches the receiver. The area serviced by a particular mast is called a **cell**. Figure 21.6 shows the role of phone masts as repeater stations and a cell network.

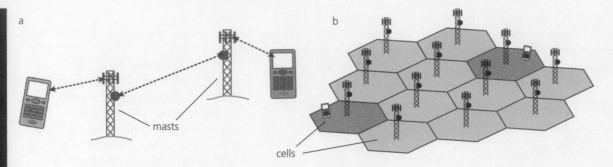

a

b

masts

cells

Figure 21.6 Repeater masts (a) and cell networks (b) in mobile phone communication

There are health risks involved in using mobile phones. It is believed that the microwaves that come from mobile phones and masts can cause harm to the body. It is suggested that mobile phones are particularly dangerous for young children because their brains are not yet fully developed.

Sensible precautions include:
- using headsets or speaker phones; this allows the phone to be kept well away from the head
- not using mobile phones for long calls (use a landline more often or text)
- siting masts well away from housing areas.

> **Exam tip**
>
> If a series of masts (repeater stations) was not used, the signals would need to be much stronger. To get a stronger signal, phones would need to be much larger.

Ⓗ Using microwaves to heat food

Microwaves cause the water molecules in food to **absorb energy**; this causes them to **vibrate more**. This rapid movement of water molecules causes the temperature of the food to rise. Because of its effect on water molecules, microwave heating is particularly effective for foods that contain a lot of water.

Dangers with electromagnetic radiation (waves)

The suggested risks with mobile phones have already been reviewed above.

Some types of electromagnetic waves (radiation) can harm living cells (living tissue). For example:
- too much UV radiation can cause skin cancer
- gamma radiation and X-rays can damage or kill cells and cause cancer.

Figure 21.5 shows that while wavelength increases from left to right, frequency increases from right to left across the range of electromagnetic waves. The waves become increasingly dangerous to human cells as their **frequency increases** and they have **more energy**; gamma rays are the most harmful to living cells.

Where possible, the risks with harmful electromagnetic radiation are being reduced. Examples include:
- using ultrasound in hospitals where possible instead of X-rays
- using X-rays in radiotherapy treatments rather than gamma radiation. (X-rays have a lower frequency and so less energy.)

> **Exam tip**
>
> Moving to the left in the electromagnetic spectrum, the frequency and energy increase. This explains why gamma radiation and X-rays can penetrate living tissue but UV radiation damages cells in the skin only.

Now test yourself

TESTED ☐

6 Name the electromagnetic wave with the longest wavelength.
7 Name the electromagnetic wave used in TV remote controls and night photography.

Answers on p. 152

Exam practice questions

1 Figure 21.7 represents some sea waves.

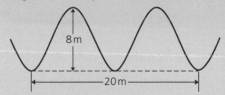

Figure 21.7

(a) What is the amplitude of these waves? [1]

(b) What is their wavelength? [1]

(c) If five complete waves pass a fixed point in 10 seconds, what is the frequency of these waves? [2]

(d) What is the unit of frequency? [1]

(e) Sea waves are transverse waves. Describe a transverse wave. [2]

2 Ultrasound travels at 1500 m/s in water. A ship sends out an ultrasound pulse towards the seabed and the echo returns 5 seconds later.

(a) Calculate the depth of the water below the ship. [2]

(b) Suggest **one** explanation for the pulse unexpectedly returning after 2 seconds. [1]

(c) The ultrasound used has a frequency of 25 000 Hz.
Use the wave equation:

speed = frequency × wavelength

to calculate the wavelength of this ultrasound. [3]

3 Figure 21.8 shows part of the electromagnetic spectrum.

gamma rays	X-rays		visible light		microwaves	radio waves

Figure 21.8

(a) Copy the diagram and add the two missing types of electromagnetic radiation. [2]

(b) Name **one** thing that all types of electromagnetic wave have in common. [1]

(c) Give **one** harmful effect of gamma radiation. [1]

4 Describe how mobile phone messages pass from person to person. [3]

Answers online

ONLINE

22 Road transport and safety

Motor vehicles must be able to stop within the available distance in front of them, otherwise a collision occurs.

Stopping a motor vehicle

Stopping distance has two parts: thinking distance and braking distance:
- Stopping distance is the distance travelled between starting to think about stopping and reaching a complete stop.
- **Thinking distance** is how far the vehicle travels while the driver is thinking about what to do.
- **Braking distance** is how far the vehicle travels after the brakes are activated until reaching a complete stop.

stopping distance = thinking distance + braking distance

Thinking distance and braking distance (and therefore stopping distance) are increased by many factors, as shown in Table 22.1.

Table 22.1 Factors that affect thinking and braking distance

Factors that increase thinking distance	Factors that increase braking distance
• Faster speed of the vehicle • Taking alcohol, drugs and medicines • Tiredness	• Faster speed of the vehicle • Poor brakes • Icy or wet weather conditions • Bald tyres
The factors above increase the distance travelled while thinking about what to do	The last two bullet points above increase the distance required to brake because they reduce friction between the tyres and the road

Reaction time

Thinking distance is affected by a driver's **reaction time**. If the reaction time is short, the thinking distance is less. Reaction time is the time that it takes someone to react to a situation, such as a driver reacting to a dog running out in front of the car. It is the time between seeing the dog and starting to do something about it.

Reaction time can be measured using a metre rule, as shown in Figure 22.1.

The steps are as follows.
1 Person **A** holds the metre rule by the tip so that it hangs vertically downwards.
2 Person **B** holds an outstretched hand against the bottom end of the metre rule in a grasping position, but not actually holding the rule.
3 Person **A** lets go of the metre rule and **B** catches it as soon as possible.
4 The distance the rule falls before being caught gives an indication of **B**'s reaction time – the longer the distance, the longer the reaction time (and slower the reaction).

> **Stopping distance**: The distance travelled between starting to think about stopping and reaching a complete stop.
>
> **Thinking distance**: The distance a vehicle travels while the driver is thinking about what to do.
>
> **Braking distance**: The distance a vehicle travels after the brakes are activated until reaching a complete stop.

> **Exam tip**
>
> Thinking distance is measured in metres (or some other unit of distance), not seconds.

> **Exam tip**
>
> If either (or both) the thinking or braking distance increases, then the stopping distance increases.

> **Reaction time**: The time it takes to react to a situation.

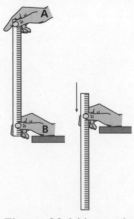

Figure 22.1 Measuring reaction time

Exam practice answers at **www.hoddereducation.co.uk/myrevisionnotesdownloads**

Friction

Braking distance depends on **friction**. Friction is a force that opposes motion (Figure 22.2). It is measured in **newtons (N)**.

← friction   motion →

Figure 22.2 Friction opposes motion

Friction is produced when two surfaces, such as the tyres of a car and the road, rub together. Factors that affect friction include weight and the type of surface.

- The heavier the weight (for example, of a car), the greater the friction.
- The rougher the surface, the greater the friction. This explains why tyres have more friction on a dry road than on an icy road.

> **Friction**: A force which opposes motion.

> **Exam tip**
>
> You need to be aware when friction can be useful and when it can be unhelpful. Friction can oppose the motion of a vehicle and, therefore, more fuel is needed to overcome the friction; this is not helpful. However, the more friction there is between the brakes and the wheels, and between the tyres and the road surface, the better for stopping.

Now test yourself

TESTED

1 Define the term 'thinking distance'.
2 Which of the values (thinking distance, braking distance and stopping distance) is/are affected by road conditions?
3 Name the force which opposes motion.

Answers on p. 152

Developing alternative fuels for transport and reducing reliance on fossil fuels

REVISED

Most cars use petrol or diesel as fuel; these are fossil fuels and so are finite resources. However, there are many ways in which manufacturers are trying to reduce the amount of fuel used by modern cars.

These include:
- stop–start technology (where the car's engine will shut down if the car is not moving)
- car parts made from lightweight materials
- better streamlining.

H Apart from the methods listed above, there are many other ways in which car manufacturers are designing cars to use less fossil fuel. These include developing cars which run on other types of fuel:
- **Biofuel** (biodiesel) can be produced from vegetable oil, such as the oil in the seeds of oilseed rape.
- **Gasohol** is produced by adding alcohol to petrol. The alcohol can be made from a range of crops including sugar beet, sugar cane, barley and potatoes.

(H) Substances that are used *instead of* petrol or diesel are called **substitutes fuels**. They includes fuels such as biodiesel/biofuel.

Substances that are used *together with* petrol or diesel are **extenders**. This includes fuels such as the alcohol in gasohol.

Other methods of reducing our reliance on fossil fuels to run our cars are detailed below.

- **Regenerative hybrid systems**: in the Toyota Prius, and some other modern cars, the high-voltage rechargeable battery is recharged by kinetic energy from moving parts in the engine. This can be so effective that the battery can remain charged and be used to power the vehicle (and save petrol).

- In **regenerative braking systems**, the kinetic energy of braking is converted to electricity by generators attached to the wheels. The electricity is used to recharge the battery.

- **Fuel cells**: these are a special type of battery that can have ingredients added (rather than the battery being recharged). There are many types of fuel that can be used in fuel cells; they include methanol. The alcohol needed to make methanol can be obtained from renewable biomass. In a fuel cell the methanol 'reacts' to produce electricity, which powers the car. Another type of fuel cell uses hydrogen as the fuel. The cell is used to produce the electricity that powers the car. A big advantage is that there is no shortage of hydrogen and its use does not pollute the atmosphere because it does not produce carbon dioxide; it releases heat and water as by-products.

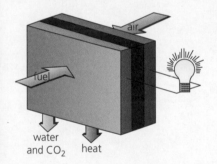

Figure 22.3 A methanol fuel cell

> **Substitute fuel**: A fuel that can be used instead of petrol or diesel, such as biodiesel.
>
> **Extender**: A substance that is added to petrol or diesel to reduce the use of fossil fuels, such as the alcohol in gasohol.

> **Regenerative hybrid system**: The use of kinetic energy from engine parts to recharge the battery in vehicles.
>
> **Regenerative braking system**: A system in which kinetic energy from braking is converted into electricity by generators linked to the vehicles' wheels.

> **Exam tip**
>
> Hybrid cars are powered by electricity *and* petrol or diesel.

> **Fuel cell**: A special type of battery that has raw materials added, such as methanol, rather than being recharged as in traditional batteries.

Road safety

REVISED

There are many safety features to improve road safety.

Safety in cars

The design of modern cars incorporates many safety features, as shown in Table 22.2.

Table 22.2 Some safety features of modern cars

Safety feature	Function
Seat belts	Restrain the driver and passengers, preventing them being thrown forward on impact
Airbags	The shock of impact causes airbags to inflate rapidly, providing a cushion between the driver (and passengers) and the steering wheel and other hard surfaces
Crumple zones	Situated at the front and rear of cars; on impact they absorb energy as they 'crumple' and collapse slowly, reducing the force that people inside the car are subjected to
Rigid passenger cells	The cabin that encloses the driver and passengers is tough and rigid; it will normally not collapse on impact, protecting those inside from crush injuries

H Safety features such as airbags and crumple zones absorb energy in collisions in order to reduce injuries to those inside a car.

Crumple zones and airbags increase the length of time that it takes for a driver and passengers to come to a complete stop, and this helps to reduce the force of the impact.

Speed restrictions (limits) and other safety measures

Speed restrictions, speed bumps and traffic cameras each play a part in contributing to road safety, as shown in Table 22.3.

Table 22.3 Speed restrictions, speed bumps and traffic cameras can reduce the number of accidents and reduce the injury to drivers and passengers if a collision occurs

Feature	Description	Function
Speed limits	These are the upper speed limits set by the government; the limits are different for different types of road – in a built-up area it is typically 30 mph, but for motorways it is 70 mph	Speed limits make accidents less likely and reduce the extent of injury if an accident occurs
Speed bumps	Slow vehicles down in built-up areas; if drivers go over bumps too quickly, they can damage their vehicles	Accidents are less likely and reduce the extent of injury if an accident happens
Traffic cameras	**Instantaneous speed cameras** measure a car's speed at a particular moment in time – the actual time the camera takes the image **Average speed cameras** measure the average speed of a car over a set distance; to measure average speed at least two cameras are required at different places along a road; these calculate the average speed by measuring the time taken for a vehicle to travel a particular distance	Traffic cameras encourage drivers to stay within speed limits as drivers can face fines or even disqualification if caught speeding

Speed

REVISED ☐

'Speed' tells us how quickly something is travelling. For example, a car could have a speed of 45 miles per hour (mph). This tells you it will travel 45 miles (distance) in 1 hour (time).

$$\text{average speed} = \frac{\text{distance travelled}}{\text{time taken}}$$

Distance–time graphs

A **distance–time graph** is a graph of distance (on the y-axis) against time (on the x-axis).

There are some key points to know when interpreting distance–time graphs:
- A straight diagonal line means that the object (for example, the car) is moving at a constant speed.
- The steeper the diagonal line, the faster the speed.
- Horizontal lines mean that the object is not moving (speed = 0).

Worked example

Figure 22.4 shows a distance–time graph for a car over a period of 60 seconds.

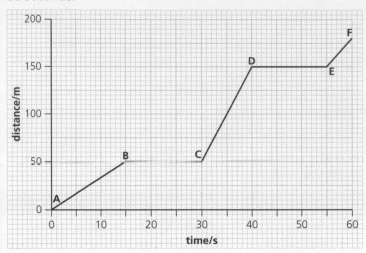

Figure 22.4 **A distance–time graph for a car journey**

(a) Use the graph to find the distance the car travelled in the first 15 seconds.

(b) Between which two letters is the car travelling fastest?

(c) Calculate the total time for which the car was stopped.

(d) Use the equation:

$$\text{average speed} = \frac{\text{total distance travelled}}{\text{total time taken}}$$

to calculate the car's average speed over the 60 seconds.

Answers

(a) 50 m

(b) **C–D** (steepest angle)

(c) **B – C + D – E** = (30 – 15) + (55 – 40) = 30 s

(d) $\frac{180}{60}$ = 3 m/s

Now test yourself

TESTED

4 How does the presence of speed bumps help road safety?

5 In a distance–time graph what does a straight horizontal line represent?

Answers on p. 152

Ⓗ Balanced and unbalanced forces

REVISED

Several forces can act on an object at any one time. For a moving car, friction with the road and the air oppose motion, the force of the engine promotes forward motion and gravity provides a downwards force.

(H) There are some key features about combined forces and motion:
- If the forces are equal in size but opposite in direction, they are **balanced**.
- If the forces are balanced, an object will either **remain at rest** or move in a **straight line with a constant speed**.
- If the forces are unbalanced, the **resultant force** causes the object to accelerate/decelerate and/or change direction.
- You should know the equation: **resultant force = mass × acceleration**

> **Resultant force**: The force produced as a consequence of unbalanced forces. A resultant force will have both a value and a direction.

Worked example

A car has a forward force of 1200 N and a backward force of 1000 N. Describe the motion of the car in terms of the resultant force.

Answer

The resultant force is 1200 – 1000 = 200 N forwards. The car accelerates in the forward direction.

Exam practice questions

1 Table 22.4 shows some information about the stopping distance of a car at different speeds.

Table 22.4

Speed/mph	Thinking distance/m	Braking distance/m	Stopping distance/m
25	7	14	21
50	15	40	
75		80	104

(a) Copy and complete the table by calculating the **two** missing values. [2]
(b) Describe the relationship between speed and stopping distance. [1]
(c) Apart from speed, state **two** factors that affect braking distance. [2]

2 (a) State **two** features of modern cars that are designed to reduce passenger injuries if a car is involved in a collision. [2]
   (b) (i) Describe how speed bumps promote driver and passenger safety. [2]
       (ii) Suggest **one** disadvantage of speed bumps. [1]

3 Figure 22.5 shows a distance–time graph for two bicycles (**A** and **B**) travelling along a road.

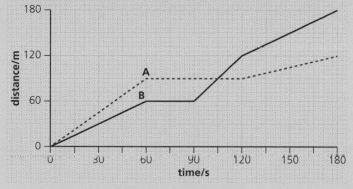

Figure 22.5 A distance–time graph for two bicycles (**A** and **B**)

(a) How far had bicycle **A** travelled after 60 seconds? [1]
(b) Which bicycle travelled the furthest over 180 seconds? [1]
(c) Which bicycle was stopped for the longest time? Explain your answer. [3]
(d) Use the equation:

$$\text{average speed} = \frac{\text{total distance travelled}}{\text{total time taken}}$$

to calculate the average speed for bicycle **A** over the entire journey. [2]

H **4** Figure 22.6 represents a car moving to the left on a straight road.

forward force
10 000 N

frictional force
10 000 N

Figure 22.6

(a) In terms of forces, describe the motion of the car. [2]

The driver increased the force on the accelerator, causing an additional forward force of 4000 N. (The frictional force was unchanged.)

(b) In terms of resultant force, describe the effect of this on the car. [3]

Answers online

ONLINE

23 Radioactivity

Types of radiation

Atoms and radiation

Matter is made up of very tiny particles called **atoms**. Inside each atom there are three types of small (subatomic) particle: **protons**, **neutrons** and **electrons**.

The nucleus of an atom contains protons and neutrons. The electrons of an atom are in orbits or electron shells around the nucleus. For each element, the number of subatomic particles in each atom is normally constant.

Radioactive elements have atoms that have unstable nuclei because of unstable combinations of protons and neutrons. These unstable atoms:
● disintegrate or decay (split up)
● form smaller, more stable atoms
● emit radiation as this happens.

Alpha, beta and gamma radiation

There are three types of radiation that can be emitted from radioactive nuclei: **alpha radiation**, **beta radiation** and **gamma radiation**. These are detailed in Table 23.1.

Table 23.1 The three types of radiation

Type	Description	Stopped by
Alpha (α)	● Large, heavy, slow particles ● Helium atom that has lost its two electrons	● A few centimetres of air ● A thin sheet of paper
Beta (β)	● Small, light and fast-moving electrons	● A few metres of air ● A few millimetres (thin sheet) of aluminium
Gamma (γ)	● Fast and powerful electromagnetic radiation	● A few centimetres of lead

> **Alpha radiation (α):** Radiation that can be stopped by a few centimetres of air or a thin sheet of paper.
>
> **Beta radiation (β):** Radiation that can be stopped by a few metres of air or a thin sheet of aluminium.
>
> **Gamma radiation (γ):** Highly penetrating electromagnetic radiation that can be stopped by several centimetres of lead.

Background radiation

Background radiation is radiation that is all around us; it is always present. As with other radiation, background radiation is caused by the disintegration of unstable nuclei.

Sources of background radiation include:
● **radon**, a radioactive gas emitted from granite rocks
● **carbon-14**, a type of carbon found inside living organisms
● **cosmic rays** reaching the Earth from space
● **nuclear reactors** in nuclear power stations.

> **Exam tip**
>
> You met gamma radiation before, in Chapter 21; it is part of the electromagnetic spectrum. You need to be aware of the fact that it has high energy and is very dangerous.

When measuring the radioactivity coming from a source, the background radiation normally contributes to the total radiation count. The actual amount of background radiation can be calculated as the amount of radiation remaining after the radiation from a source reduces to zero.

Uses of radiation

Radioactivity can be used in industry, medicine and agriculture.

Gamma rays can be used to:
- **kill cancer cells** in the process of radiotherapy
- **sterilise** surgical instruments by killing any microorganisms (bacteria, fungi or viruses)
- **preserve** (extend the shelf-life of) perishable fresh food by killing microorganisms and stopping decay
- **detect leaks** from underground pipes; if engineers suspect an underground pipe has a leak they can add a radioactive substance to the pipe and monitor radioactive levels in the ground immediately above the track of the pipe along the ground.

Now test yourself

TESTED

1 Name the three types of radiation.
2 Name the type of radiation that can be stopped by a few metres of air or a few millimetres of aluminium.
3 What name is given to the type of radiation which is all around us?

Answers on p. 152

Half-life

REVISED

In a radioactive element, atoms decay and disintegrate, so the number of atoms of that radioactive element must decrease with time.

The **half-life** is the length of time it takes for half of the atoms in a radioactive sample to disintegrate. In effect, the level of radioactivity falls by half.

Figure 23.1 shows what happens during a period of three half-lives of a radioactive element.

> **Half-life**: The time taken for the level of radioactivity from a source to fall by half.

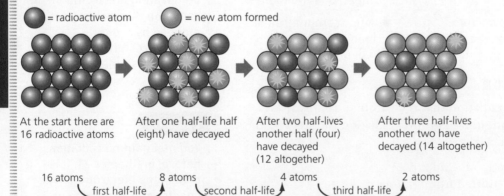

| At the start there are 16 radioactive atoms | After one half-life half (eight) have decayed | After two half-lives another half (four) have decayed (12 altogether) | After three half-lives another two have decayed (14 altogether) |

16 atoms → 8 atoms → 4 atoms → 2 atoms
first half-life   second half-life   third half-life

Figure 23.1 Half-life

Half-life can be represented by a graph, as shown in Figure 23.2.

> **Exam tip**
>
> In exam questions, background radiation is often the explanation used to account for graphs showing the amount of radiation levelling off but not reaching zero.

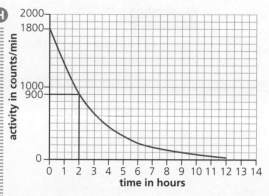

Figure 23.2 Calculating half-life from a graph

In Figure 23.2 it takes 2 hours for the activity (level of radiation) to fall from 1800 counts/minute to 900 counts/minute. Therefore the half-life of this element is 2 hours. After 4 hours, the activity is 450 counts/minute and so on.

> **Exam tip**
>
> Many exam questions ask about linking the half-life of radioactive elements with particular roles in medicine or industry. In these questions, you have to work out whether a long half-life or a short half-life (or something in between) is most useful.

Worked example

A radioactive element gives a reading of 80 counts/minute. Two hours later the reading was 20 counts/minute. Calculate the half-life of this element.

Answer

1 hour

20 is a quarter of 80 (after one half-life the value fell to 40, and then to 20 after the second half-life); so 2 hours represents two half-lives.

Ionising radiation

REVISED

(Normal) atoms are electrically neutral because they have equal numbers of (positive) protons and (negative) electrons. When alpha, beta or gamma radiation collides with atoms, electrons tend to be displaced from their shells. As a result, the atoms become positively charged because they now have more protons than electrons.

These charged atoms are now **ions**; the process of electrically neutral atoms becoming ions is called **ionisation**.

Alpha and beta particles and gamma radiation are called **ionising radiations** because they cause ionisation.

> **Ionisation**: The process of the formation of ions by radiation.
>
> **Ionising radiation**: Radiation that causes ionisation (the formation of ions).

Risks with using ionising radiation

Although ionising radiation (gamma rays) is used in the treatment of cancer during radiotherapy (Figure 23.3), gamma radiation can be very dangerous as it kills cells in the body.

The amount (dose) of radiation used must be calculated carefully and targeted accurately on cancer cells to ensure that the cancer cells are killed and that as few as possible normal (healthy) cells are damaged.

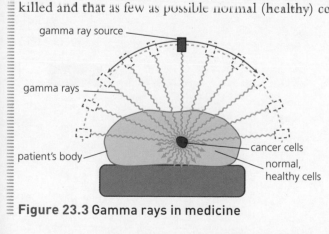

Figure 23.3 Gamma rays in medicine

Whenever ionising radiation is used in industry, medicine or agriculture, measures are put in place to reduce the possible harm caused. This is why patients with cancer will usually get their radiotherapy treatment in a series of very small doses. Figure 23.3 shows how changing the angle of penetration of the different doses can reduce the harm caused to healthy cells in the body. Safety concerns are also the reason why the radiographers leave the room during the time the radiation is actually administered (similar to dentists leaving the room when X-rays of teeth are taken).

Now test yourself

TESTED

4 What is meant by the term half-life?
5 Give **one** use of gamma radiation in hospitals.

Answers on p. 152

Exam practice questions

1 Explain why some elements are radioactive. [2]
2 (a) Explain what is meant by 'background radiation'. [1]
  (b) Give **one** source of background radiation. [1]
3 A radioactive source and a radioactive detector are placed as shown in Figure 23.4.

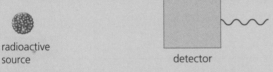

radioactive source

detector

Figure 23.4

(a) Describe how you could determine whether the radioactive source emits alpha, beta or gamma radiation. [4]
(b) State **two** things that would have to be kept the same to make the results valid. [2]
(c) State why the count never reaches zero. [1]
4 The activity of a radioactive isotope changes as shown in Table 23.2.

Table 23.2

Time/days	Activity/cpm
0	440
1	330
2	220
3	165
4	110
5	82

(a) Describe the trend shown by the table. [1]
(b) What is the half-life of the isotope? [1]

Answers online

ONLINE

24 Earth in space

The Solar System

REVISED

Our Solar System has one **star** (the Sun) surrounded by eight **planets** that travel in paths called **orbits**.

Figure 24.1 shows the Sun, the eight planets and their orbits.

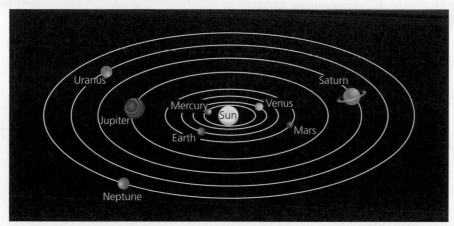

Figure 24.1 The Solar System

> **Star**: A massive extra-terrestrial structure that emits light and radiation.
>
> **Planet**: A structure that is in orbit around a star.
>
> **Orbit**: The circular/elliptical path of a structure travelling around a star or a planet.

Some key features of the Solar System:
- All the planets travel around the Sun in the same direction.
- The further a planet is from the Sun, the longer it takes to orbit it.
- The planets further from the Sun have elliptical (oval-shaped) orbits; those closer (such as Mercury, Venus, Earth and Mars) have more circular orbits.
- The planets closer to the Sun have a solid surface and are rocky planets (Mercury, Venus, Earth and Mars). Those further away are mainly formed of gas and are called the gas planets.

Some of the other structures in the Solar System are listed below:
- **Comets** are formed from rock covered by frozen water and frozen gases; these orbit the Sun.
- **Asteroids** are large chunks of rock that also orbit the Sun. Most asteroids are found in a zone (the asteroid belt) between Mars and Jupiter. Occasionally, an asteroid can be knocked out of its orbit and come close to the Earth. Collisions between asteroids and the Earth are possible and have happened in the past. Evidence includes the existence of large **craters** caused by such impacts.
- **Moons** orbit planets. The Earth has one moon; there are 173 moons altogether around the planets in the Solar System.
- **Artificial satellites** are also orbiting some of the planets; most are in orbit around Earth. These have been sent into space by man for a number of functions, including observing the Earth (such as military satellites), weather monitoring, astronomy (studying the stars) and communications (mobile phones and computers).

> **Exam tip**
>
> You need to know the order of the planets moving away from the Sun: they are Mercury, Venus, Earth, Mars, Jupiter, Saturn, Uranus and Neptune.

> **Exam tip**
>
> You should be aware that the further a planet is from the Sun, the colder it is likely to be.

> **Comet**: Ice-covered rocky structure that orbits the Sun.
>
> **Asteroid**: A very large rock found in space.
>
> **Moon**: A (natural) structure that orbits a planet.

Gravity

Gravity is the force of attraction that exists between objects.

The force of gravity depends on:
- the **mass** of the objects; the bigger the objects, the larger the force of gravity
- the **distance** between the objects; the force of gravity decreases as the distance apart increases.

Gravity can be very important; it is the force that keeps planets, comets, moons and artificial satellites in orbit. If there was no gravity, they would all float out into space.

The force of gravity is what causes all objects to have **weight**. Gravity is measured in **newtons per kilogram** (**N/kg**). On the Earth's surface, an object of mass 1 kg has a weight of 10 N.

Weight can be calculated using the equation:

weight = mass × gravity

Worked example

The Earth is about six times larger than the Earth's moon. Explain why astronauts landing on the Moon can jump large distances, even when wearing very heavy spacesuits.

Answer

As the Moon is smaller, its gravity is much lower (about six times less) than that of the Earth.

Now test yourself

TESTED

1 Name the planet that is third closest to the Sun.
2 State the relationship between a planet's distance from the Sun and the time it takes to orbit the Sun.
3 Give **one** function of artificial satellites.

Answers on p. 152

⊕Stars and galaxies

REVISED

Stars

All **stars** are formed by the same sequence of events.
1 A cloud of **hydrogen** comes together under the force of gravity.
2 As the hydrogen is compressed (pulled together), becoming very dense; its temperature rises to many million degrees.
3 At these extremely high temperatures, **nuclear fusion** reactions occur and the hydrogen nuclei combine to form heavier nuclei such as helium.
4 This nuclear fusion emits light and heat (and other radiation).

Galaxies

A **galaxy** is a huge collection of stars held together by gravity. The galaxy that contains our Solar System is called the **Milky Way**.

Gravity: The force of attraction that exists between objects.

Exam tip

You should know that your mass (in kg) will not change wherever you are in the Universe, but that your weight (in N) would be different on different planets due to the effect of the size of the different planets on gravity. Humans in space weigh less than those on the Earth's surface as the distance between them and Earth is much greater.

Nuclear fusion: The joining together of two or more (lighter) nuclei (e.g. hydrogen) to form a heavier nucleus (e.g. helium) with the release of a lot of energy.

Exam tip

Nuclear fusion is the process by which stars gain their energy.

Galaxy: A huge collection of stars held together by gravity.

Exam practice answers at **www.hoddereducation.co.uk/myrevisionnotesdownloads**

(H) The **Universe** is the name given to the space occupied by all the galaxies that exist.

Galaxies do not remain settled in one position in the Universe; they move away from each other.

The distances between stars (and between different galaxies) are so large that we do not use the normal units of measurement, such as kilometres. Distances between stars and galaxies are measured in **light years**. One light year is the distance that light travels in 1 year.

The expanding Universe

REVISED

There are two key features about the movement of galaxies.
- They are continually moving away from each other and the further away from each other they are, the faster they are moving apart.
- As galaxies move apart, space and the Universe are expanding.

All the stars in the Universe emit light in the form of electromagnetic radiation. Analysis of the spectra of this starlight, which ranges from violet to red on the basis of wavelength, shows the presence of black lines (**absorption spectra**) where atoms of hydrogen (and other elements) have absorbed some of the light.

When we examine the light spectra from stars in distant galaxies, we find that:
- the absorption spectrum pattern is broadly the same across all galaxies
- the pattern is shifted (in Figure 24.2) towards the red end of the spectrum; this is called **red-shift**
- the further away a galaxy, the larger the red-shift.

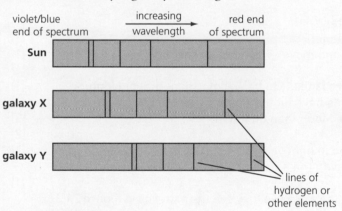

Figure 24.2 Red-shift in distant galaxies

The red-shift happens because the light coming from a source that is moving away from us has a longer wavelength than it would have if the source was stationary.

Worked example

Use Figure 24.2 to explain how you know that galaxy **Y** is further away from us than galaxy **X**.

Answer

The light from galaxy **Y** has a bigger red-shift than the light from galaxy **X**; equivalent lines have moved further to the red end of the spectrum. The bigger the red shift, the longer the distance.

Universe: The space occupied by all the galaxies that exist.

Light year: The distance that light travels in 1 year.

Exam tip

Remember that a light year is a distance, not a time; it is approximately 9.4×10^{12} kilometres.

Exam tip

You should appreciate the size of space: a typical galaxy can have over a billion stars and the Universe contains more than a hundred billion galaxies!

Red-shift: The phenomenon of the light spectrum of distant galaxies being shifted towards the red end of the spectrum.

The origin of the Universe

Most scientists think that the **Big Bang theory** is the best explanation for the origin of the Universe.

The Universe started as a tiny point, called a **singularity**, about **14 billion years ago**. The Big Bang represents a massive cosmic expansion (explosion) that expelled matter and energy in all directions out from this central point. Over millions of years, gravity pulled the scattered matter together to form galaxies, stars, planets and moons. The forces involved are causing the Universe and space to continue to expand; evidence of this is provided by the red–shift of distant galaxies.

There are other theories as to how the Universe began; the best known is the **Steady State theory**. This proposes that the Universe has not changed much throughout time and that it had no significant beginning, such as the Big Bang. It suggests that new matter is continually created as the Universe expands. Most scientists do not support this theory.

Now test yourself

TESTED

4 Name our galaxy.
5 What name is given to the space occupied by all the galaxies that exist?
6 Define the term 'light year'.

Answers on p. 152

Exam practice questions

1 Copy and complete the following sentences.
   (a) _____ are objects that orbit a star. [1]
   (b) _____ are objects that orbit a planet. [1]
   (c) _____ contain millions of stars and move away from each other. [1]
2 (a) Name the planet closest to our Sun. [1]
   (b) Suggest why the surface temperature of Mars is likely to be warmer than that of Neptune. [1]
   (c) What is the evidence that shows that asteroids have hit the Earth in the past? [1]
3 Explain fully why a human's weight is less on the Moon than on Earth. [3]
4 (a) What is meant by the term 'light year'? [2]
   (b) Explain why distances in space are measured in light years. [1]
5 (a) Write the following in order of size, starting with the smallest:
      Universe   Solar System   galaxy   star. [1]
   (b) When scientists analyse the light from two distant galaxies they see the patterns in Figure 24.3.

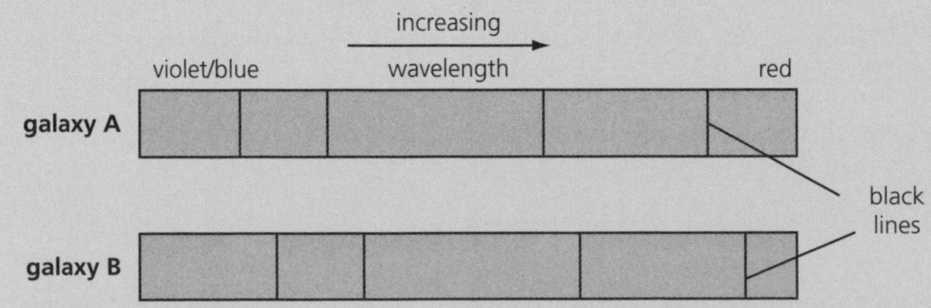

Figure 24.3

   (i)   What causes the difference in the patterns? [1]
   (ii) State **one** conclusion that can be drawn from this information. [1]
   (iii) Suggest how light coming from the Milky Way would be different from the patterns shown in the diagram. Explain your answer. [3]

Answers online

ONLINE

Glossary

Biology

Abiotic factor: A non-living factor, such as CO_2 level, that can be used to monitor environmental change.

Active immunity: The type of immunity produced by the body making antibodies.

Allele: One of two possible versions of a particular gene.

Amniocentesis: A process in which foetal cells are obtained from the amniotic fluid and then examined for the presence of genetic abnormalities.

Amnion: A lining produced during pregnancy that contains the amniotic fluid.

Amniotic fluid: The fluid within the amnion that cushions the foetus.

Antibiotic: A chemical produced by fungi that kills bacteria.

Antibiotic resistance: Antibiotic-resistant bacteria cannot be killed by at least one type of antibiotic.

Antibody: A structure produced by lymphocytes that has a complementary shape (and can attach) to antigens on a particular microorganism.

Antigen: A distinctive marker on a microorganism that leads to the body producing specific antibodies.

Association neurone: A neurone that connects sensory and motor neurones.

Auxin: The plant hormone involved in the phototropic response.

Benedict's test: The food test for a (reducing) sugar.

Binge drinking: Drinking a lot of alcohol over a short period of time.

Biodiversity: The range/number of species in an area.

Biotic factor: A living organism, such as lichen, that can be used to monitor environmental change.

Biuret test: The food test for protein.

Bronchitis: The narrowing of the airways in the lungs; usually caused by smoking tobacco.

Brownfield site: A site that has previously had housing, or other urban structures, on it.

Cancer: Uncontrolled cell division.

Cardiac output: The volume (amount) of blood the heart pumps per minute.

Cell: The basic building block of all living organisms.

Cell membrane: The selectively permeable boundary to plant and animal cells.

Cell wall: A rigid structure immediately outside the cell membrane in plants that provides support. Plant cell walls are formed of cellulose.

Central nervous system (CNS): The part of the nervous system that links receptors and effectors, comprising the brain and the spinal cord.

Cervix: The opening of the uterus.

Chloroplast: A structure in a plant that contains chlorophyll. Photosynthesis in plants takes place in chloroplasts.

Cholesterol: A fatty substance that causes narrowing of the blood vessels.

Chromosome: A genetic structure that occurs in functional pairs in the nucleus of cells (except gametes, where there is only one chromosome from each pair).

Circulatory system: The body system that includes the heart and blood vessels.

Communicable disease: A disease that can be passed from one organism (person) to another.

Competitive invasive species: A non-native species (introduced by man) that spreads rapidly, outcompeting native species.

Complication: A long-term health problem (e.g. of diabetes).

Condom: A barrier contraceptive method.

Continuous variation: The type of variation in which there is a gradual change in a feature, with no distinct groups.

Contraception: A method used to try to avoid pregnancy.

Contraceptive pill: A pill containing hormones that prevents pregnancy by affecting hormone levels, thereby preventing eggs being released.

Coronary arteries: The blood vessels that supply the heart with blood.

Cystic fibrosis: A genetic condition in humans caused by having two recessive alleles of a particular gene.

Cytoplasm: The part of a cell between the cell membrane and the nucleus. Chemical reactions in the cell take place here.

Diabetes: A condition in which the blood glucose control mechanism fails.

Diploid: Describes a cell or nucleus with the normal chromosome number.

Discontinuous variation: The type of variation in which all the individuals can be clearly divided into a small number of groups with no overlap.

DNA: The molecule (core component) that forms genes and chromosomes.

Dominant: In the heterozygous condition, the dominant allele overrides the recessive allele.

Double helix: The structure of DNA.

Down's Syndrome: A genetic condition in humans caused by having one extra (47) chromosome.

Effector: Muscles (and glands) that can produce a response when stimulated.

Egg (ovum): A female gamete.

Emphysema: Damage to the gas exchange surfaces in the lungs; usually caused by smoking tobacco.

Endothermic reaction: A reaction that requires energy to be absorbed (taken in) to work.

Ethanol: The food test for fats/oils.

Evolution: The continuing process of natural selection that leads to the change in a species over time (or the formation of a new species).

Exothermic: A chemical reaction which releases energy (heat).

Extinction: A species is extinct if there are no living members of that species left.

Female sterilisation: A contraceptive method in which the oviducts are cut.

Fertilisation: The joining together of a sperm and an egg to produce a zygote.

Food chain: The order in which energy passes through a sequence of living organisms.

Food web: A number of interlinked food chains.

Fossil: The remains of a living organism that has been preserved (usually in rock) for millions of years.

Gamete: Sex cell that contains only one chromosome from each pair.

Gene: A short section of a chromosome that codes for a particular characteristic.

Genetic condition: A condition caused by problems in genes or chromosomes.

Genetic engineering: A process which modifies the genome of an organism.

Genetic screening: A process used to test people for the presence of particular harmful alleles or other genetic abnormalities.

Genome: The entire genetic (DNA) make-up of an individual.

Genotype: The genetic make-up of an individual represented by two symbols (letters)

Glycogen: A storage compound found in the liver consisting of glucose sub-units.

Haploid: Describes a cell or nucleus with half the normal number of chromosomes.

Heterozygous: The two alleles of a particular gene are different.

Homozygous: Both alleles of a particular gene are the same.

Hormone: A chemical messenger produced by a gland that travels in the blood to a target organ where it acts.

Implantation: The term describing the attachment of the ball of cells (embryo) following fertilisation to the uterus lining.

Insulin: The hormone that lowers glucose levels in the blood.

***In vitro* testing**: The initial stage of testing drugs and medicines. It involves testing the drugs on cells and tissues in the laboratory.

Iodine: The food test for starch.

Leukaemia: A type of cancer in which some types of blood cells increase in number and cell division is out of control.

Lymphocyte: A type of white blood cell that produces antibodies.

Memory lymphocyte: A special type of lymphocyte that can remain in the body for many years and produce antibodies quickly when required.

Menstrual cycle: The monthly cycle in females of reproductive age that prepares the body for pregnancy.

Menstruation: The breakdown and removal of the blood-rich uterine lining at the end of each cycle.

Motor neurone: A neurone that carries impulses from the CNS to effectors.

Multicelled organism: An organism that is formed of many cells.

Mutation: A random change in the structure of a gene or number of chromosomes.

Natural selection: The process in which the better adapted individuals survive (at the expense of the less well adapted ones) long enough to reproduce and pass on their genes to their offspring.

Nerve cell (neurone): A specialised cell found in the nervous system. Nerve cells conduct nerve impulses.

Nerve impulse: An electrical signal that travels through nerve cells (neurones).

Nucleus: The control centre of the cell

Oestrogen: The female hormone that both causes the repair and build-up of the uterine lining following menstruation and stimulates ovulation.

Organ: A structure in the body formed of different tissues to carry out a particular function.

Organ system: Organs are organised into organ systems.

Ovary: A female organ that produces eggs (ova).

Oviduct: Carries eggs (ova) from the ovary to the uterus.

Ovulation: The release of an ovum (egg) by an ovary.

Passive immunity: The type of immunity produced by injecting antibodies.

Pedigree diagram: A diagram that shows how a particular condition is inherited through the different generations of a family.

Penicillin: The first antibiotic developed.

Penis: Organ that introduces sperm into the vagina

Phagocyte: A type of white blood cell that destroys microorganisms by engulfing them and then digesting them (phagocytosis).

Phenotype: The outward appearance of a feature.

Photosynthesis: A process in plants in which light energy is trapped by chlorophyll to produce sugars and starch (food).

Phototropism: A plant growth response which results in plant stems growing in the direction of a light source.

Placenta. The structure that links the uterus wall to the foetus via the umbilical cord. It is here that exchange of materials takes place between the mother and the foetus.

Primary consumer: An animal that eats plants.

Producer: A plant that makes food by photosynthesis. Producers always make up the first stage in a food chain.

Progesterone: The female hormone that maintains the build-up of the uterus lining and prepares the uterus for pregnancy.

Prostate gland: Adds fluid to feed the sperm.

Punnett square: A grid (table) used to work out the offspring in a genetic cross.

Receptor: A part of the body that, upon stimulation, can cause an impulse to be sent to the CNS.

Recessive allele: An allele that will only show in the phenotype if both alleles are recessive (and there is no dominant allele present).

Reflex action: A rapid, involuntary action that does not involve conscious thought.

Reflex arc: The pathway of neurones in a reflex.

Respiration: The process that releases energy from food in all cells.

Scrotum: Sac that holds and protects the testes.

Secondary (or tertiary) consumer: An animal that eats other animals.

Sensory neurone: A neurone that carries nerve impulses from a receptor to the CNS.

Side effect: An unwanted or unplanned effect of a drug on a person.

Sperm: A male gamete (sex cell).

Sperm tube: Tube that carries sperm from the testes to the penis.

Starch test: A test to show whether or not starch is present in a plant leaf.

Stem cell: A simple cell in plants and animals which has the ability to divide to form cells of the same (general) type.

Stroke: A circulatory disease that affects the brain.

'Superbug': A type of bacterium that is resistant to a number of antibiotics.

Symptom: A sign that shows that something is medically wrong.

Synapse: A small gap between neurones.

Testis: Produces sperm.

Tissue: A group of cells with similar structures and functions.

Urethra: Tube through which the sperm leaves the penis.

Uterus: Place where the foetus will develop if pregnancy occurs.

Vacuole: A large fluid-filled structure in plant cells that contains sap.

Vagina: Opening of the female reproductive system into which sperm is deposited during sexual intercourse.

Vasectomy: Male sterilisation. A contraceptive method in which the sperm tubes are cut.

Voluntary action: An action or response that involves conscious thought.

Zygote: The first (diploid) cell of the new individual following fertilisation.

Chemistry

Acid: A solution with a pH less than 7.

Addition polymerisation: A reaction in which small molecules (monomers) are joined or added together to make a long chain molecule (a polymer).

Alkali: A solution with a pH more than 7.

Alkali metal: A metal in group 1 of the Periodic Table.

Alkaline earth metal: A metal in group 2 of the Periodic Table.

Alkanes: A group (homologous series) of saturated hydrocarbon compounds with the general formula C_nH_{2n+2}.

Alkenes: A group (homologous series) of unsaturated hydrocarbon compounds with the general formula C_nH_{2n}.

Alternative light source: A non-visible light source, e.g. UV light, which can be used to help visualise a fingerprint.

Anion: A negative ion.

Anode: The positive electrode in electrolysis.

Antacid: A medicine used to treat acid indigestion.

Atom: The smallest sub-unit of an element; the smallest part of an element that can exist. Atoms are made up of even smaller sub-units called protons, neutrons and electrons.

Atomic number: The number of protons in the nucleus of an atom.

Base: An insoluble alkali, with a pH more than 7.

Biodegradable: A substance that can be decomposed (broken down) by microorganisms.

Boiling: The process during which a liquid changes into a gas.

Boiling point: The temperature at which a liquid turns into a gas.

Catalyst: A substance which causes a reaction to go faster without being used up during the reaction.

Cathode: The negative electrode in electrolysis.

Cation: A positive ion.

Chemical developer: A dye that can be used to help visualise a fingerprint.

Chromatogram: The visualisation of results of chromatography.

Chromatography: A separation technique used to separate the different soluble substances in a mixture, based on their different solubilities.

Combustion: The reaction of a fuel with oxygen, commonly called burning.

Complete combustion: The burning of a fuel when there is a good supply of oxygen.

Compound: A substance that has the atoms of two or more different elements chemical bonded (joined) together.

Condenser (Liebig): The apparatus used to turn a gas into a liquid by cooling.

Condensing: The process during which a gas turns into a liquid.

Covalent bond: A shared pair of electrons that holds two atoms together.

Covalent compound: Two or more atoms joined together by sharing electrons (a covalent bond).

Crude oil: A fossil fuel; a liquid mixture containing many different substances that can be separated by fractional distillation.

Crystallisation: The process used to separate a dissolved solid from a solvent.

Diatomic: A molecule containing two atoms.

Distillate: The solvent from a solution that has been separated by the process of distillation.

Distillation: The process used to separate a solvent from a solution. Distillation involves both evaporation and condensation.

Electrode: The piece of apparatus, usually graphite rods, used to conduct electricity into the electrolyte.

Electrolysis: The process of using electricity to decompose an ionic compound.

Electrolyte: The liquid that conducts electricity.

Electron: The negative subatomic particle found in the electron shell(s) of an atom.

Electron shell: The zone(s) around the nucleus of an atom in which electrons are located.

Electronic configuration (structure): The arrangement of electrons in the electron shells of an atom.

Element: A substance which contains only one type of atom.

Endothermic: A reaction in which heat energy is taken in.

Exothermic: A reaction that gives out heat energy.

Filtrate: The liquid that passes through the filter funnel during filtration.

Exam practice answers at **www.hoddereducation.co.uk/myrevisionnotesdownloads**

Filtration: The process used to separate an insoluble solid from a liquid.

Fingerprint: The unique pattern of lines found on a person's fingertips.

Finite: A defined amount of a resource that cannot be replaced when used.

Flame test: A test used to identify some metal ions by the flame colour they produce when burning.

Fraction: A mixture of molecules with similar boiling points.

Fractional distillation: A method used to separate miscible liquids with different boiling points. The process used to separate the compounds in crude oil. It involves heating and evaporation, and the subsequent condensation of gases back to liquids.

Freezing: The process during which a liquid changes into a solid.

Fullerene: A family of carbon molecules each with carbon atoms linked in rings to form a hollow sphere or tube.

General formula: The formula of an homologous series that shows the relationship between the number of carbon atoms in the molecule and the other elements.

Global warming: An increase in the temperature of the Earth's surface.

Greenhouse effect: The heating effect caused by a layer of greenhouse gases trapping heat within the Earth's atmosphere.

Group: A vertical column in the Periodic Table. The group number gives the number of electrons in the outer shell of the elements in that group.

Halogens: The elements in group 7 of the Periodic Table.

Hazard symbol: A symbol used on a chemical to warn of danger.

Homologous series: A group of compounds with the same general formula and similar chemical properties.

Hydrocarbon: A compound containing carbon and hydrogen only.

Indicator: A chemical that can change colour to show whether a substance is an acid, neutral or an alkali.

Indigestion: The effect caused by excess acid in the stomach.

Inert: Unreactive.

Insoluble: A substance that will not dissolve in a solvent.

Ion: A charged particle produced as a result of an atom gaining or losing one or more electrons.

Ionic compound: A compound formed between a metal and a non-metal in which the metal atom transfers electron(s) to the non-metal atom.

Leading edge: The top of a spot on a chromatogram.

Limewater: The chemical used to test for the presence of carbon dioxide.

Litmus indicator: An indicator made from lichens.

Lone pair: A pair of electrons not involved in bonding.

Mass number: The total number of protons and neutrons in the nucleus of an atom.

Melting: The process during which a solid changes into a liquid.

Melting point: The temperature at which a solid turns into a liquid.

Miscible: Miscible liquids can mix with other miscible liquids.

Mixture: Two or more substances together that are not chemically bonded (joined).

Mobile phase: The liquid that travels up the stationary phase during chromatography.

Monomer: A small molecule that can be joined together with many other molecules to make a polymer.

Nanomaterial: A material made from nanoparticles.

Nanoparticles: Structures that contain a few hundred atoms; typically 1–100 nm in size.

Natural material: A material that can be obtained from living things, such as wool, or made without being processed by chemical methods, such as granite.

Neutralisation: The reaction between an alkali (base) and an acid. Also refers to the process of using alkali to 'neutralise' the effect of acid, for example in indigestion remedies.

Neutron: An electrically neutral subatomic particle found in the nucleus of an atom.

Noble gases: The elements (gases) in group 0 of the Periodic Table.

Non-biodegradable: A substance that cannot be decomposed (broken down) by microorganisms.

Nucleus: The central part of an atom containing the protons and neutrons.

Organic chemistry: The study of carbon-containing compounds

Period: A horizontal row in the Periodic Table. The period number gives the number of electron shells in the atoms of the elements in that period.

Periodic Table: The table that lists the known elements in order of atomic number.

pH scale: The scale used to measure the strengths of acids and bases.

Photochromic: The term given to paints and dyes (or other materials) that change their colour as light intensity changes.

Polymer: A large molecule produced by the joining together of many monomers in a process called polymerisation.

Polymerisation: The process in which a long chain molecule is made from joining lots of small molecules (monomers) together.

Proton: The positive subatomic particle found in the nucleus of an atom.

Pure: A single element or compound that is not mixed with any other substance.

Rate of reaction: The amount of reactant used or product formed per unit time.

Raw material: The starting material for a manufacturing process.

Reactivity series: The order of metals according to how reactive they are with substances such as water or acid.

Residue: The solid left in filter paper after filtration or in the evaporation dish during evaporation.

R_f value: The value given to the ratio of how far a substance travels during chromatography compared with how far the solvent travels.

Salt: The substance formed during an acid reaction.

Simple distillation: The process of evaporation and condensation used to separate a mixture.

Smart material: A material that changes its properties if exposed to a particular environmental change.

Soluble: Refers to a substance that will dissolve in a solvent.

Solute: The solid that dissolves in a solvent.

Solution: The mixture of soluble solid (solute) and liquid (solvent).

Solvent: The liquid that dissolves the solute (soluble solid).

Solvent front: The distance the solvent travels through the stationary phase during chromatography.

State: The physical state of a substance; whether it is solid, liquid or gas.

Stationary phase: In chromatography, the phase that allows the solvent to travel though it; it is used to separate the components in the mixture.

Sublimation: The process during which gases change directly into solids or solids change directly into gases.

Synthetic material: A man-made or manufactured material, for example plastic.

Thermochromic: The term given to paints or dyes that change their colour when their temperature changes.

Trace evidence: A small amount of evidence left at a crime scene, such as hair or fibres from clothing. Trace evidence normally needs scientific analysis.

Transition metals: The block of elements between group 2 and group 3 of the Periodic Table.

Universal indicator: An indicator that can change to a number of different colours when added to a neutral substance or different strengths of acids or alkalis.

Physics

Airbag: Safety device in a vehicle that inflates rapidly upon impact. Airbags serve to cushion the driver/passenger from hard objects, such as the steering wheel, in a collision.

Alpha radiation (α): Radiation that can be stopped by a few centimetres of air or a thin sheet of paper.

Ammeter: An instrument for measuring electrical current.

Ampere (amp): The unit of electrical current (represented by the symbol A).

Amplitude: The maximum height of a wave.

Asteroid: A very large rock found in space.

Audible hearing range: The range of frequencies that humans can hear (20 Hz to 20 kHz).

Average speed: The average speed that a vehicle is travelling at over a set distance.

Background radiation: Radiation that is always around us.

Balanced forces: Forces that are both equal in size and opposite in direction.

Battery: Two or more electrical cells.

Beta radiation (β): Radiation that can be stopped by a few metres of air or a thin sheet of aluminium.

Big Bang theory: The theory explaining the origin of the Universe. The Big Bang proposes that the Universe originated as a tiny point, a singularity, 14 billion years ago, and has continued to expand ever since.

Biofuel: Fuel produced from (recently) living material.

Blue (coloured) wire: The neutral wire in a three-pin plug.

Boiler: The part of a fossil fuel power station that converts water into steam. The boiler converts thermal energy into kinetic energy.

Braking distance: The distance a vehicle travels after the brakes are activated until reaching a complete stop.

Brown (coloured) wire: The live wire in a three-pin plug.

Cell polarity: The existence of positive and negative terminals in a cell or battery.

Circuit diagram: A diagram showing the wiring and components in an electrical circuit.

Comet: An ice-covered rocky structure that orbits the Sun.

Conduction: A process that describes the transfer of heat (thermal) energy through the particles of a solid. Materials, such as metals, that conduct heat easily are said to be good conductors of heat energy.

Conductor: 1 A material that allows electricity to flow through it easily. 2 A material (e.g. metal) that conducts heat easily.

Convection: The transfer of heat energy in a liquid or a gas.

Conventional current: The imagined flow of electricity from the positive terminal of a cell or battery to the negative terminal through a circuit.

Crumple zone: A region in a vehicle that 'crumples' and collapses slowly upon impact in a collision so reducing the force that people inside the vehicle are subjected to.

Current: The amount of electricity flowing around a circuit (or through a component).

Distance–time graph: A graph of distance (on the y-axis) against time (on the x-axis).

Double insulation: A safety system which encloses all conducting parts of an electrical circuit in a plastic box, so that the user can never touch a live component and get an electrical shock.

Dynamo: A device for converting the kinetic energy of a moving object into electricity.

Echo: Reflected sound.

Efficiency: A measure of how much of the input energy (to a process or device) is converted to useful output energy.

Electrical cell: A component that supplies electricity.

Electrical (national) grid: The system of pylons and cabling that distributes electricity from power stations around the country.

Electricity: The flow of electrons through a conductor.

Electromagnetic spectrum: The range of electromagnetic waves in order of increasing wavelength and decreasing frequency (gamma rays – X-rays – ultraviolet – visible light – infrared – microwaves – radio waves)

Electron: 1 A subatomic particle in an atom that has a negligible relative mass and a relative charge of −1. Electrons orbit the atom nucleus in zones called shells. 2 Negatively charged particles which can move around an electrical circuit conducting electricity; particles that can conduct heat in metals.

Extender: A substance that is added to petrol or diesel to reduce the use of fossil fuels, such as the alcohol in gasohol.

Flash-bang method: A method of measuring the speed of sound in air.

Fossil fuel: Fuel formed from the remains of plants and animals that have been compressed by layers of rock for millions of years.

Frequency: The number of waves passing a particular point in one second.

Friction: A force which opposes motion.

Fuel cell: A special type of battery that has raw materials added, such as methanol, rather than being recharged, as in traditional batteries.

Fuse: A safety device consisting of a fine wire which melts if too much current flows through it, thus breaking an electrical circuit.

Galaxy: A huge collection of stars held together by gravity.

Gamma radiation (γ): Highly penetrating electromagnetic radiation that can be stopped by several centimetres of lead.

Gasohol: Fuel produced by adding alcohol to petrol.

Generator: The part of a power station that makes electricity (similar in principle to a dynamo) by converting kinetic energy to electrical energy

Gravitational field strength: A measure of how strong the force of gravity is. On Earth, it is about 10 N/kg.

Gravitational potential energy: The energy possessed by an object because of its height above the ground.

Gravity: The force of attraction that exists between objects.

Green and yellow (coloured) wire: The earth wire in a three-pin plug.

Half-life: The time taken for the level of radioactivity from a source to fall by half.

Hertz: The unit of frequency, represented by the symbol Hz.

Instantaneous speed: Speed at a particular moment in time.

Insulator: 1 A material that does not allow electricity to flow through it easily. 2 A material that does not transfer heat energy easily.

Ionisation: The process of the formation of ions by radiation

Ionising radiation: Radiation that causes ionisation (the formation of ions).

Joule: The unit of energy, represented by the symbol J.

Kilowatt-hour: A 'unit' of electricity, represented by the symbol kWh. It is the amount of electricity used by an appliance that uses 1000 W of power for 1 hour.

Kinetic energy: The energy possessed by an object due to its movement.

Light year: The distance that light travels in one year.

Longitudinal wave: A wave in which the particles vibrate in the same direction as (parallel to) the direction of travel.

Mobile phone cell: The area serviced by a particular mast.

Moon: A (natural) structure that orbits a planet.

Neutron: A subatomic particle in an atom's nucleus that has a relative mass of 1 and a relative charge of 0.

Newton (N): The unit of force, e.g. friction.

Non-renewable energy: Energy from a source which has a finite supply (it will eventually run out).

Nuclear fusion: The joining together of two or more (lighter) nuclei (e.g. hydrogen) to form a heavier nucleus (e.g. helium) with the release of a lot of energy.

Ohm: The unit of resistance (represented by the symbol Ω).

Ohm's law: When electricity flows through a metal wire at constant temperature, the voltage and current are directly proportional to each other.

Orbit: The circular/elliptical path of a structure travelling around a star or a planet.

Parallel circuit: A circuit which has more than one branch through which electricity can flow.

Planet: A structure that is in orbit around a star.

Power: The rate at which electrical energy is transferred or work is done.

Principle of the Conservation of Energy: Energy can be changed from one form to another but the total amount of energy does not change.

Proton: A subatomic particle in an atom's nucleus that has a relative mass of 1 and a relative charge of +1.

Radiation: 1 Radiation is emitted by radioactive elements as their nuclei disintegrate. 2 The transfer of heat energy by waves passing out from a hot surface.

Radioactive element: An element that has atoms with unstable nuclei due to unstable combinations of protons and neutrons.

Reaction time: The time it takes to react to a situation.

Red-shift: The phenomenon of the light spectrum of distant galaxies being shifted towards the red end of the spectrum.

Regenerative braking system: A system in which kinetic energy from braking is converted into electricity by generators linked to the vehicles' wheels.

Regenerative hybrid system: The use of kinetic energy from engine parts to recharge the battery in vehicles.

Renewable energy: Energy collected from sources that will never run out or from sources that are naturally replenished within a human lifetime.

Resistance: The opposition by a material to the flow of electrical current.

Resistor: A device that can limit and change the size of electrical current.

Resultant force: The force produced as a consequence of unbalanced forces. A resultant force will have both a value and a direction.

Rigid passenger cell: The strong rigid cabin that encloses the driver and passengers in a vehicle. It will normally remain intact in a collision preventing those inside from getting crushed.

Series circuit: An electrical circuit in which the components are connected in sequence (one after another or side by side).

Solar System: The Sun and the structures (e.g. planets, comets and asteroids) that orbit it.

Speed: How fast an object is travelling. Speed can be measured in metres/second (m/s), kilometres/hour (km/hr) or miles per hour (mph).

Speed bump: A deliberately placed ridge in the road that serves to reduce vehicle speed.

Speed limit: The maximum legal vehicle speed in a particular area.

Star: A massive extraterrestrial structure that emits light and radiation.

Steady State theory: An alternative theory (largely discarded) to the Big Bang theory for the origin of the Universe.

Step-down transformer: Device that decreases electrical voltage and increases the current as electricity leaves the electrical grid and is used by businesses and homes.

Step-up transformer: Device that increases voltage and decreases the current of electricity before it flows through the electrical grid.

Stopping distance: The distance travelled between starting to think about stopping and reaching a complete stop.

Substitute fuel: A fuel that can be used instead of petrol or diesel, such as biodiesel.

Thinking distance: The distance a vehicle travels while the driver is thinking about what to do.

Traffic calming measure: A strategy that helps reduce vehicle speed in a particular area, e.g. road narrowing schemes.

Transformer: A device which converts high voltages to low voltages and vice versa.

Transverse wave: A wave in which the particles vibrate at right angles to the direction of wave travel.

Turbine: A machine (in a power station) that drives the generator (converts the kinetic energy of the turbine into the kinetic energy of the generator). The blades of the turbine turn due to the kinetic energy of the steam.

Ultrasound: Sound waves with a frequency higher than 20 000 Hz (20 kHz)

'Unit' of electricity: An 'amount' of electricity that is used when calculating electricity bills.

Universe: The space occupied by all the galaxies that exist.

Variable resistor (rheostat): A device that can change the current flowing through a component, e.g. a dimmer light switch.

Volt: The unit of voltage (represented by the symbol V).

Voltage: The amount of electrical energy supplied to a circuit; it is the voltage that causes an electrical current to flow.

Voltmeter: An instrument for measuring voltage.

Watt (kilowatt): The unit of power, represented by the symbol W (kW).

Wavelength: The distance between two successive wave crests or troughs.

Weight: The force of gravity on an object.

Now test yourself answers

Chapter 1 Cells

1 Cellulose cell wall, permanent vacuole, chloroplasts
2 Cell wall
3 The coverslip protects the objective lens (should it come in contact with the slide) and stops the cells from drying out.
4 Leukaemia
5 Groups of cells with similar structures and functions

Chapter 2 Food and diet

1 Carbohydrates
2 Starch
3 **glucose + oxygen → carbon dioxide + water + energy**
4 Stroke
5 Increasing exercise, reducing stress, stopping smoking

Chapter 3 Chromosomes and genes

1 Nucleus
2 Double helix
3 Down's Syndrome
4 The two alleles of a gene are the same.
5 A recessive allele will not appear in the phenotype unless both alleles are recessive.
6 One
7 Genetic screening is a process used to test people for the presence of particular harmful alleles or other genetic abnormalities.
8 It can produce insulin or other products of use.

Chapter 4 Coordination and control

1 Brain and spinal cord
2 A rapid, involuntary action that does not involve conscious thought.
3 Association neurone
4 Pancreas
5 Glucose is converted to glycogen and glucose is moved from blood into cells where it is used during the process of respiration.

6 Any **three** from:
   ○ high blood glucose
   ○ glucose in the urine
   ○ lethargy
   ○ thirst.

Chapter 5 Reproductive system

1 Feed (nourish) the sperm
2 Oviduct or Fallopian tube
3 Cushions the foetus
4 Sperm and ova (eggs)
5 The breakdown and removal of the blood-rich uterine lining at the end of each menstrual cycle.

Chapter 6 Variation and adaptation

1 Genetic and environmental
2 Continuous and discontinuous
3 A continuing process of natural selection which leads to gradual change in organisms over time and which may result in the formation of new species.
4 Fossils can show that species have changed and also the way in which they have changed (the intermediate steps).

Chapter 7 Disease and body defences

1 A disease that can be passed from person (organism) to person (organism).
2 Bacteria
3 Lymphocyte
4 Fast acting and short lived
5 A chemical produced by fungi that kills bacteria.
6 Fleming, Florey and Chain
7 *In vitro* testing, animal testing, clinical testing and licensing

Chapter 8 Ecological relationships

1 **carbon dioxide + water → glucose + oxygen**
2 A leaf that is part green and part white (or other colour)/a leaf that only has chlorophyll in some places.

3 Producers
4 Any **two** from:
  ○ water
  ○ food
  ○ territory
  ○ mates.
5 A non-native species (introduced by man) that spreads rapidly, outcompeting native species.
6 A non-living factor, e.g. CO_2, that can be used to monitor the environment.
7 The number/range of species in an area.
8 It creates more habitats.

Chapter 9 Acids, bases and salts

1 Ethanoic acid/citric acid
2 Beetroot/red cabbage/blackcurrant/other appropriate example
3 Neutralisation (involving an acid and a base/alkali)
4 Salt and hydrogen

Chapter 10 Elements, compounds and mixtures

1 Freezing
2 The process during which a gas turns into a liquid.
3 A substance that has two or more different elements chemically joined (bonded) together.
4 Two (H and Cl)
5 A solid that can dissolve in a liquid (solvent).
6 Filtration

Chapter 11 Periodic Table, atomic structure and bonding

1 Protons, neutrons and electrons
2 The total number of protons and neutrons in an atom (nucleus).
3 Alkaline earth metal
4 Two
5 Metal hydroxide and hydrogen
6 A charged particle produced when an atom gains or loses one or more electrons.
7 Covalent

Chapter 12 Metals and the reactivity series

1 sodium – calcium – aluminium – copper
2 **aluminium + sulfuric acid → aluminium sulfate + hydrogen**
3 They move to the cathode and gain electrons.
4 **$Al^{3+} + 3e^- → Al$**

Chapter 13 Materials

1 Cheaper/better properties
2 Plastic/glass/any appropriate example
3 Thermochromic
4 Evidence that can be obtained from digital devices, e.g. computers and mobile phones.
5 Arch, loop, whorl and composite

Chapter 14 Rates of reaction

1 Reactant; product
2 The reaction is getting faster.

Chapter 15 Organic chemistry

1 C_nH_{2n+2}
2 Surfacing roads and roofs
3 There is a double covalent bond between two carbon atoms.
4 C_nH_{2n}
5 **alkane/alkene + oxygen → carbon dioxide + water**
6 Ethene
7 The monomer (ethene) has a double bond/there are no double bonds in polythene

Chapter 16 Electrical circuits

1 Measure voltage
2 A material that allows electricity to flow through it easily.
3 Ampere (amp)
4 An electrical circuit in which the components are arranged in sequence (one after another).
5 Shared
6 Shared

Chapter 17 Household electricity

1 Brown
2 Returns electricity to the plug socket from an appliance.
3 Watt/kilowatt
4 The amount of electricity used by an appliance of 1000 W in 1 hour.
5 kWh

Chapter 18 Energy

1 Chemical
2 Electrical → sound
3 Energy can be changed from one form to another, but the total amount of energy does not change.

Chapter 19 Electricity generation

1 Turbine
2 Kinetic energy to electrical energy
3 Increases voltage and decreases current
4 Supplies of uranium ore will eventually run out.
5 Destroys habitats/expensive to build

Chapter 20 Heat transfer

1 A material that does not transfer heat energy easily
2 Convection
3 Radiation
4 Electrons
5 The gas particles in hot air move further apart and the hot air becomes less dense, causing it to rise.

Chapter 21 Waves

1 Sound/ultrasound
2 The maximum height of a wave
3 Hertz (Hz)
4 20 Hz–20 kHz
5 Sound waves with a frequency higher than 20 000 Hz
6 Radio waves
7 Infrared

Chapter 22 Road transport and safety

1 How far a vehicle travels while the driver is thinking about what to do.
2 Braking and stopping distance
3 Friction
4 Slows down vehicles in built-up areas, making accidents less likely to happen. If they do happen, the accidents are normally less serious.
5 Vehicle stopped

Chapter 23 Radioactivity

1 Alpha, beta and gamma
2 Beta
3 Background radiation
4 The time for the level of radioactivity to fall by half.
5 Radiotherapy/to kill cancer cells/to sterilise surgical instruments

Chapter 24 Earth in space

1 Earth
2 The further a planet is away from the Sun, the longer it takes to orbit the Sun.
3 Weather monitoring/communications/astronomy/military functions/other appropriate answer
4 Milky Way
5 The Universe
6 The distance that light travels in one year.

Exam practice answers at **www.hoddereducation.co.uk/myrevisionnotesdownloads**